Patents for Chemical Inventions

Symposia sponsored by the Division of Chemical Literature and the Division of Industrial and Engineering Chemistry at the 145th Meeting of the American Chemical Society, New York, N. Y. September 9 and 13, 1963

Elmer J. Lawson and

Edmund A. Godula, *Symposia Chairmen*

ADVANCES IN CHEMISTRY SERIES **46**

AMERICAN CHEMICAL SOCIETY

WASHINGTON, D. C.   **1964**

Copyright © 1964

American Chemical Society

All Rights Reserved

Library of Congress Catalog Card 64-24274

PRINTED IN THE UNITED STATES OF AMERICA

Advances in Chemistry Series

Robert F. Gould, *Editor*

Advisory Board

Fred Basolo

Raymond F. Boyer

John H. Fletcher

Jack Halpern

Wayne W. Hilty

George W. Irving

Amel R. Menotti

Walter C. Saeman

Leo H. Sommer

AMERICAN CHEMICAL SOCIETY

APPLIED PUBLICATIONS

FOREWORD

ADVANCES IN CHEMISTRY SERIES was founded in 1949 by the American Chemical Society as an outlet for symposia and collections of data in special areas of topical interest that could not be accommodated in the Society's journals. It provides a medium for symposia that would otherwise be fragmented, their papers distributed among several journals or not published at all. Papers are reviewed critically according to ACS editorial standards and receive the careful attention and processing characteristic of ACS publications.

PREFACE

By coincidence, during the week of September 9-13, 1963, at the 145th Meeting of the American Chemical Society in New York City two symposia on patents were scheduled and held. We, as the organizers, were unaware of each other's efforts, and so it was a very satisfying piece of good fortune that the topics selected for the two symposia were such that they were complementary, rather than overlapping.

The Monday symposium, chaired by Edmund A. Godula and presented before the Division of Industrial and Engineering Chemistry, covered the broad range of problems involved in getting a patent, without too much specific concern as to the nature of the invention, these problems being generally concerned with the nature of invention, ownership, signatory formalities during prosecution, inventorship, and priority.

By contrast, the Friday symposium, chaired by Elmer J. Lawson, and presented before the Division of Chemical Literature, opened with a general discussion of the requirements for patentability, i.e., novelty, utility and unobviousness, then concentrated on specific constantly recurring chemical situations where the subject matter was structurally related to prior art, or involved a "new use," or "natural product," and concluded with a discussion of how patent protection possibilities on chemicals and medicinals vary in different countries.

Collecting these symposium papers in this volume offered an opportunity to include also a paper by Sidney G. Berry entitled "Chemical Patents, Their Meaning and Interpretation," as presented before the Columbia University School of Library Science "Institute on Patents as a Source of Information" in June 1960.

These papers, taken all together, admittedly do not cover the full range of ways patents concern chemists and others. The use of patents as literature, the drafting of patent claims, detailed procedural aspects of patent prosecution, problems of validity and infringement, the law and tactics of patent and invention licensing—these subjects and many other aspects of patent lore have been slighted or omitted. Such topics, although of obvious importance, are of greater interest to specialists. It is hoped this volume will give a fair, if non-specialist, concept of "chemical patent law" to research chemists and managers, and to them—the sources of most chemical inventions—this volume is dedicated.

EDMUND A. GODULA
Parker and Carter
Chicago, Illinois

ELMER J. LAWSON
Sterling-Winthrop
Research Institute
Rensselaer, New York

CONTENTS

Foreword — iv

Preface — v

Legal Abbreviations Used — viii

1. **Legal Requirements for Patentability** — 1
   Joseph Schimmel, U. S. Patent Office, Washington, D. C.

2. **Patentability in Chemical Inventions** — 7
   Michael G. Berkman, Kegan, Kegan & Berkman, Chicago, Ill.

3. **Inventorship in Chemical Patents** — 15
   Edmund A. Godula, Parker and Carter, Chicago, Ill.

4. **Ownership in Inventions** — 27
   J. R. Janes, Watson, Leavenworth, Kelton & Taggart, New York, N. Y.

5. **Priority of Invention** — 41
   Robert L. Niblack, Abbott Laboratories, North Chicago, Ill.

6. **Formal Documents of the U. S. Patent System** — 49
   Paul D. Burgauer, Abbott Laboratories, North Chicago, Ill.

7. **Meaning and Interpretation of Chemical Patents** — 57
   Sydney G. Berry, Berry and Crews, New York, N. Y.

8. **Patentability of Homologs, Isomers, and Other Analogs** — 73
   Dean Laurence, Laurence and Laurence, Washington, D. C.

9. **Patent Protection Available on New Uses for Old Chemicals** — 81
   S. Branch Walker, American Cyanamid Co., Stamford, Conn.

10. **Patentability of Natural Products, Plant Isolates, and Microbiological Products** — 99
    John H. Schneider, Johnson and Johnson, New Brunswick, N. J.

11. **Foreign Patent Coverage on Chemicals and Medicinals** — 107
    Alan Swabey, Alan Swabey & Co., Montreal, Canada

Legal Abbreviations Used

C.C.P.A. = Court of Customs and Patent Appeals or published reports of its decisions, U.S. Government Patent Office, Washington, D. C.
C. Ct. = Circuit Court, as "Cir. Ct. Ill.," now U.S. District Court for the District of Illinois
Cir. = U.S. Court of Appeals, usually with circuit No., as 7th Cir.
Ct. App. = Court of Appeals
Ct. Cl. = Court of Claims
Ct. Cust. App. = Court of Customs Appeals (Reports)
C.D. = Commissioners' Decisions. Decisions made by the U.S. Commissioner of Patents, published annually by the U.S. Government Printing Office, Washington, D. C.
D. Ct. = U.S. District Court, as "D. Ct. Md." = U.S. District Court of Maryland; or "D.Ct. N.D.Ill." = U.S. District Court, Northern District of Illinois.
Ex parte = In the interest of
Fed. = *Federal Reporter*, West Publishing Co., St. Paul, Minn.
F. 2d = *Federal Reporter*, 2nd series, West Publishing Co., St. Paul,, Minn.
F. Supp. = *Federal Supplement*, West Publishing Co., St. Paul, Minn.
How. = Howard (U.S. 1843-1860)
In re = In the matter of
J.O.P.S. = *Journal of the Patent Office Society*, U.S. Patent Office
L. Ed. = Lawyer's Edition, U.S. Supreme Court Reports
M.P.E.P. = "Manual of Patent Examining Procedure," U.S. Patent Office
O.G. = *Official Gazette of the U.S. Patent Office*
P.O. Bd. App. = U.S. Patent Office Board of Appeals
P.O. Bd. Interfer. = U.S. Patent Office Board of Interferences
Sup. Ct. = U.S. Supreme Court
Supra = above
U.S. = Official edition of the decisions of the U.S. Supreme Court
U.S.C. = U.S. Code
U.S.C.A. = U.S. Code annotated
U.S.P.Q. = *U.S. Patents Quarterly*, Bureau of National Affairs, Washington, D. C.
v. = versus

1

Legal Requirements for Patentability

JOSEPH SCHIMMEL, Deputy Solicitor
U.S. Patent Office, Washington, D. C.

Patents are granted only for inventions in machines, methods, manufactures, compositions of matter, designs, and asexually produced plants. An invention, to be patentable, must be novel, useful, and unobvious. The requirement of novelty means that something new is disclosed. Novelty exists if the invention has not been described in a publication or patent anywhere in the world, or if there has been no prior public use or sale in this country. The requirement of utility means the performance of some beneficial function, though only crudely. The requirement for unobviousness means that the subject matter would not be obvious to a skilled worker, if he had available to him all prior knowledge. All three requirements are essential, and all three must coexist.

The belief held by many that an inventor has an exclusive right in his invention is a fallacy. Independent of statute law, no such right exists. While an inventor may enjoy that right as long as he keeps his invention secret, the moment he discloses it by making and selling that which he invented, he abandons his exclusive right. From that moment others acquire the right to make, use, and sell the invention. Hence, exclusivity granted by patent depends entirely upon statute, and, in general, that exclusivity may be said to be the reward granted the inventor for disclosing his invention to the public. The reward usually takes the form of certain exclusive privileges for a limited period of time.

Historical Background

Historically, patents for inventions were granted by many of the

American colonies. The first of such patents, we are told, was in the field of chemistry—a method of making salt—granted to Samuel Winslow in 1641 by the Massachusetts Colony. One of the most famous of such patents was that granted in 1787 to John Fitch by New York for the sole and exclusive right to make and use boats propelled by fire or steam. This patent was later repealed, then granted to Robert Livingston. But when his efforts were fruitless, the patent was extended, in 1803, jointly to Livingston and Fulton, when the latter successfully launched the *Clermont*. The first patent granted under the seal of the United States was also a chemical patent—a method of making potash and pearl ash—granted to Samuel Hopkins on July 31, 1790.

Now, considering the federal patent system, it is of the utmost importance to realize that the keystone and foundation therefor is the provision in Clause 8 of Section 8 of Article I of the Constitution, which contains a broad grant of power to Congress "to promote the progress of science and useful arts by securing for limited periods to . . . . inventors the exclusive right to their . . . . discoveries." No restraints are placed on the power of Congress to legislate in this field. Hence, Congress has complete authority and may modify the statutory law at will, as long as rights of property in existing patents are not affected. Furthermore, the primary and paramount purpose of the constitutional grant is "to promote the progress of science and useful arts," and reward to the inventor is the means devised to accomplish that objective. Hence, the public good takes precedence over the principle of reward, the latter being of secondary consideration and incidental to the public interest. Because of this it has been said that the legal requirement for patentability—that is, the standard of patentability—is a constitutional standard.

The Constitution speaks of "inventors" and "their discoveries." The primary and common meaning of "discoveries" is not synonymous with inventions, since, as pointed out by Webster, discoveries bring to light that which existed before, but was unknown, whereas invention applies to the production of something that did not exist before. As used in the Constitution and patent laws, "discovery" and "discoveries" are always considered as synonymous with invention. While every invention may involve a discovery, every discovery is not an invention. A classic example of a nonpatentable discovery was that of the use of ether by Dr. Morton, whose patent was held invalid more than 100 years ago.

Subject Matter

In accordance with the power conferred by the Constitution and in pursuance of the objective stated therein, Congress from time to time enacted laws relating to patents. The first patent act enacted in 1790 authorized the grant of a patent for any new and useful art, machine, or manufacture, provided that certain conditions existed. In 1793, the act was amended to include compositions of matter in the specification of subject matter entitled to patent protection. In 1902, patents were authorized for designs; and in 1930, plants, asexually reproduced, were included in the categories of things for which patents could be granted. Although Congress has enlarged the fields in which patents are granted, the basic fundamental legal requirements for patentability, which were set forth in the first patent act of 1790, have remained substantially unchanged throughout the years, and even the law presently in force—the Patent Act of 1952—was characterized by its Congressional sponsors as essentially no more than a codification of prior laws with certain minor changes.

Since patents are products of laws passed by the Congress, it is important to understand that patent protection can only be obtained for those classes of invention which are particularly specified in the law. The law currently in force authorizes the grant of patents for matter falling into six classes or categories—namely, arts, machines, manufactures, compositions of matter, designs, and certain types of plants. Though Congress has said that anyone who makes a new and useful invention in any of these fields may obtain a patent therefor, the granting of a patent has always been surrounded by certain formal requirements and conditions. This basic right or privilege is set forth in the present law in Section 101, and except for certain special situations it can only be exercised by the inventor himself, if living, or his legal representative, if the inventor dies before applying for the patent. One of the most important of those conditions dates back to the first patent act of 1790. It embodies a fundamental concept in order to ensure that the Constitutional objective will be attained. That concept, set forth in the requirements of Section 112 of the present act, may be stated simply as this:

In return for a full and complete disclosure to the public of a new and useful invention, the government will grant to the inventor the right to exclude others from making, using, and selling the invention for a limited period—namely 17 years.

But even this simple statement of the fundamental principle of our patent laws, which has been a guiding principle for almost two centuries, has presented and still presents many problems. What is meant by new? What is meant by useful? What is an invention? And what constitutes a full and complete disclosure? These are problems of great concern and difficulty not only to inventors but to the Patent Office in the day-to-day administration of its duty and obligation to examine applications for patents to determine whether patents are legally warranted and also to the federal courts which must, when an inventor tries to exercise the right of exclusivity granted by the patent, pass on the question of whether such patents as are issued have been properly and legally granted.

Legal Requirements

The legal requirements for patentability are threefold—novelty, utility, and nonobvious subject matter. All three must coexist to justify the grant of a patent. Standards by which novelty and unobviousness may be determined are set forth in the law, but no standards have been established for determining utility or usefulness.

Novelty

Newness or novelty of subject matter in the prescribed statutory classes has always been a requirement of the law. On the question of newness, Section 102 of the present act states that a patent may be granted, unless the invention was known or used by others in this country, and not patented or described in any printed publication anywhere before the invention thereof by the particular person applying for the patent thereon, or more than one year prior to the date of application for said patent. Thus, many things may be new within the meaning of the term "in the law" in addition to those things which are really new in the ordinary meaning of the word. For example, prior knowledge or use in a foreign country of the subject matter in question could still be considered new in the sense of our law. Even prior knowledge by some person in this country of such knowledge and use abroad does not affect the statutory definition of novelty. To negative "novelty," according to the law, such prior knowledge or use must be described in a patent or other printed publication.

However, prior knowledge or use in this country, by even a single person other than the applicant for patent, may completely destroy any claim of novelty, according to our law. However, there are recognized exceptions to this rule, particularly where the prior use was merely experimental or where the use had been abandoned or forgotten. The latter doctrine—forgotten or lost art—is applicable primarily to arts or processes, because of the intangible nature thereof, but the doctrine is inapplicable to machines, manufactures, or compositions of matter, which still exist in their entireties, since they can be identified and analyzed if necessary.

Printed publications and patents of prior date describing the invention negative a claim of novelty. Generally speaking, the former term means anything which is printed or otherwise reproduced and made available to any part of the public in any country. However, a paper read before a scientific body would not be within the prohibition of the statutory language. And prior knowledge, use, patents, and publications, all must contain information that is precise and complete enough to enable any person skilled in the art to which it relates to understand and reproduce what is disclosed therein in order to negative the novelty requirement of the law and thus defeat a subsequent inventor's right to a patent.

Utility

Since the Constitutional provision provides for the promotion of the "useful" arts, Congress has always restricted the right to obtain a patent to one whose invention is "useful." Hence, utility, like novelty, is an essential legal prerequisite to patentability. No standards or definitions of usefulness were ever established by Congress. What standards have been set are those set by judicial interpretation. Broadly speaking, it has been said that an invention is useful within the meaning of the statute, if it is capable of performing some beneficial function claimed for it, and a function is said to be beneficial as long as it is not frivolous or injurious to the well being, health, or sound morals of society. Under this definition it would seem to be clear that useless articles and all inventions which cannot accomplish or perform their specific intended functions are not within the ambit of the statute. If this definition be applied to inventions in chemical compounds and compositions of matter it would appear to be questionable whether patents for such inventions are legally justified or warranted, unless it is clear that they are capable of performing or being used for at least one specific beneficial function. The mere production of a new chemical compound or composition is not enough. The capability of such compound or composition of producing a beneficial function must be known and stated to meet the statutory requirement for utility. It is not necessary, however, that the function be performed perfectly. Capacity to do the thing intended, though only in a crude way, is one of the acid tests of utility, as a legal requirement for patentability.

Unobviousness

Patents are legally grantable only for things invented and not for things otherwise produced. The statute provides that things to be patented must be invented, as well as new and useful. Prior to the present patent act the courts had generally concluded that the word "invention" cannot be defined in such a manner as to afford any substantial aid in determining whether any particular subject matter involves invention—namely, an exercise of the inventive faculty as distinguished from simple mechanical skill. Because of this, the Act of 1952 included a section (Section 103) which was an attempt to define the term "invention" as a guide or standard, in the effort to lend substantial aid to those whose duty it is to make that determination. In the language of that section, assuming that the invention is novel in the sense previously discussed, a patent may not be obtained if the subject matter as a whole would have been obvious to a worker of ordinary skill in the art at the time the invention was made. Hence, even if novelty and utility are present, patentability depends now upon "unobviousness" instead of "inventive ingenuity" or "uncommon skill."

The question of obviousness or unobviousness, like the prior question of invention, is a question of fact and is determinable only on the basis of the evidence in each case. Moreover, as technology advances and the skill of the art shifts, the issue necessarily depends upon a shifting rather than a stated standard. Hence, there is no absolute rule today by which to determine the presence or absence of this legal requirement. However, this does not mean that the criteria or standards previously used and considered important to the determination of this question are inapplicable under the present statutory requirement. Such factors as economy, efficiency, progress of the art, long-felt want, long experimentation, general adoption by the art, specific utility, and other functions or advantages may be considered evidence of unobviousness. A statement made by Judge Bradley in 1882 has become a classic. It states that it was never the object of the patent laws to grant a monopoly for every shadow of a shade of an idea which would naturally occur to any skilled person in the ordinary course of his work. And it is still applicable. In the chemical field this principle and the prior criteria find frequent application in inventions involving homologs or analogs of known compounds; and in inventions in compositions of matter, where, unless new and beneficial results are shown for combinations of old compounds—what has been denominated synergism—the chances are good that the combination will be considered obvious, rather than unobvious, and, hence, unpatentable. In any event whether we speak or think in terms of unobviousness or mere skill, the law requires something in addition to novelty and utility. Thus, novelty and utility alone are not enough. Novelty and unobviousness also are not enough. No two of the three basic legal requirements suffices. All three requirements are essential; and all three must coexist to meet the legal requirements for patentability.

RECEIVED October 10, 1963.

2

Patentability in Chemical Inventions

MICHAEL G. BERKMAN

Kegan, Kegan & Berkman, 79 West Monroe Street, Chicago, Ill.

Whether a given invention in the chemical arts is "obvious," in the patent sense, or whether it is not is a question of increasing perplexity and interest. Another obstacle confronting the aspirant to a patent is the requirement that he establish "utility" for the invention claimed. In Patent Office practice this requirement is almost uniquely restricted to chemical and allied fields since in other areas of the "useful" arts, it is assumed that utility exists. The development and evolution of the problems posed and their current treatment are indicated in recent decisions of the Patent Office and of the courts. This article comprises an objective analysis of the most significant recent developments in these important areas of chemical patent prosecution.

The requisites for the validity of a patent (2) are novelty, utility, and invention. This article is directed to an interpretation of these requirements. It must be recognized that the standard of patentability is a constitutionally required one (12), and that the question of the validity of a patent is a question of law (20, 31). This question is to be resolved by considering whether or not the patent exhibits what is called "invention." The elusive concept of invention does not lend itself to affirmative definition, but several reasonably reliable tests have been established, consistent with the premise that the function of a patent is to add to the sum of useful knowledge (47).

The specific requirement of utility stems from the Constitutional Mandate of Article I, Section 8, and has been present in every Patent Act since the Act of 1790. The authority granted Congress is limited by the express requirement that the power be used "to promote the progress of . . . useful arts."

Under Section 101 of 35 U.S.C. of the Patent Act of 1952:

"Whoever invents or discovers any new and useful process, machine, manufacture, or composition of matter, or any new and useful improvement thereof, may obtain a patent therefor . . ."

That is, to be patentable, an invention must be useful. Section 112 of 35 U.S.C. provides:

"The specification shall contain a written description of the invention, and of the manner and process of making and using it, in such full, clear, concise, and exact terms as to enable any person skilled in the art . . . to make and use the same . . ."

There must be a disclosure of "how to use."

Utility Requirement in Chemical and Pharmaceutical Cases

Patentable utility problems are almost uniquely peculiar to chemical and pharmaceutical cases. In the mechanical and electrical fields, the useful purpose of the invention is often obvious or readily apparent, and in these areas there is ordinarily little real distinction between the discovery of a property and the development of a structural embodiment to accomplish an ultimate use.

In an important 1950 case (*5*), the rule propounded was:

". . . that no 'hard and fast' ruling properly may be made fixing the extent of the disclosure of utility necessary in an application, but . . . *the law requires an assertion of utility and an indication of the use or uses intended.*"

In a more recent decision (*34*) bearing on the problem of patentable utility in chemical applications, the court was faced with the problem of determining whether a chemical intermediate, for which no final commercially useful product is specified, has patentable utility—that is, whether usefulness as an intermediate is a patentable utility within the purview of the statute. The allegations and disclosure were found to meet the Bremner rule—but the court modified that rule by holding that "an assertion of utility" is a "meaningless formality" and not required by law. Application of the Rule of Bremner was further limited in another case (*48*), in which the court reviewed that rule and concluded there was in fact no statutory basis for such a rule, particularly where processes are involved.

The standard of 35 U.S.C. 112—"how to use"—is a factual standard, but the standard changes with time (*38, 48*). "A specification (*46*) which teaches those skilled in the art to use the process, i.e., by disclosing the manipulative steps of the process, the required operating conditions, and the starting materials so that the process may be used by a person skilled in the art, meets the requirements of 35 U.S.C. 112." It has been held not necessary to specify the intended use for the product produced therein. In general, however, the product must be shown to have some utility either discovered by the inventor or known to the prior art (*4, 22, 35*).

For many years the Patent Office has followed the practice of requiring a particularly high degree of proof of utility in cases involving therapeutic products (*28, 41*). The type and amount of test data required to establish the utility of new drugs will depend upon the particular circumstances of each case (*13, 21*). But the Patent Office has required proof of successful tests on humans whenever it may be inferred from the specification that therapeutic use in humans is contemplated. Tests on experimental animals have been considered insufficient.

Several recent appellate decisions (*8, 25*), are significant in that they establish that an inventor may comply with the statutory utility mandate without being obliged to prove an unspecified ultimate utility rather than a claimed immediate use and in that they support the premise that the interim results of organized research can be of sufficient importance to warrant patent protection. These cases, and others (*14, 36*), focus attention on the importance of the form of the utility allegation and compel the inference that the nature of the allegation will prescribe what the applicant may be called upon to prove and the nature and degree of proof required.

Applicants and their attorneys prosecuting chemical and pharmaceutical applications will find the way less arduous and the likelihood of ultimate success greatly increased if they are able to build adequate records so as to obviate Section 101 and 112 issues. A definitive statement of the use (the patentable utility) of the invention, where not obvious, should be included in the specification. Uses for which only meager support can be mustered should not be included. The intended use having been carefully defined, there should be an adequate detailed disclosure of how to use the product, method, or process of the invention.

Requirement of "Unobviousness"

The "utility" requirements having been satisfied, it is necessary to establish that the invention represents a significant advance and is not merely an "obvious" development devoid of patentable novelty.

Section 103 of 35 U.S.C., bearing on the question of "obviousness," is a new section consistent with decisions of the courts (since 1950) holding patents invalid on the ground of lack of invention or lack of patentable novelty (*6*). Earlier cases (*11, 19, 24*), in which the requirement of "invention" was stated, had the underlying concept that the presence of invention was to be determined as of the time the invention was made. This concept is now expressed in 35 U.S.C. 103 (*37*), a statutory version of what was, prior to Jan. 1, 1953, the judge-made (or case law) requirement of invention (*38*).

In the language of the 1952 statute:

"A patent may not be obtained though the invention is not identically disclosed . . . if the difference between the subject matter sought to be patented and the prior art are such that the subject matter as a whole would have been obvious at the time the invention was made to a person having ordinary skill in the art . . ."

In principle (38), "the Section 103 requirement of unobviousness is no different in chemical cases than with respect to other categories of patentable inventions."

A recently reported case (37) presented for determination the question of whether what is obvious to one of ordinary skill in the art is to be determined as of "the time the invention was made," as specified in 35 U.S.C. 103, or whether it is to be determined as of some later date when the application is filed. In analyzing the problem and reversing the lower tribunal, the court concluded (44) that ". . . 35 U.S.C. 103 is very specific in requiring that a rejection on the grounds the invention 'would have been obvious' must be based on a comparison of the prior art and the subject matter as a whole *at the time the invention was made.*"

"Hindsight" is not a proper basis for determining patentability. Although a particular solution to a problem may seem simple in retrospect after its disclosure and thereby permit experts to correlate the prior art piecemeal to the invention, foresight applied as of the date of the invention is the only proper test of invention (1).

Homologs and Obviousness—Unexpected Properties

There is no area of patent application prosecution in which rejections for "obviousness" have been asserted more automatically than in cases involving chemical compounds which are homologs or "near homologs" of a known compound. Under recent decisions of the courts, while there may be a presumption of obviousness in such cases, the presumption is rebuttable, and a clear showing of different and unexpected properties for the "homolog" may establish its patentability.

A statement of the Court of Customs and Patent Appeals in a recent case (9, 38) is particularly significant:

"From the standpoint of patent law, a compound and all its properties are inseparable; they are one and the same thing. The graphic formulae, the chemical nomenclature, the systems of classification and study such as the concepts of homology, isomerism, etc., are mere symbols by which compounds can be identified, classified, and compared. But a formula is not a compound and while it may serve in a claim to *identify* what is being patented, . . . the *thing* that is patented is not the formula but the compound identified by it. And the patentability of the thing does not depend on the similarity of its formula to that of another compound but on the similarity of the former compound to the latter. There is no basis in law for ignoring any property in making such a comparison. An assumed similarity based on a comparison of formulae must give way to evidence that the assumption is erroneous."

In another decision (*40*), the same court said:

While "... homology provides for the chemist a convenient system of structural classification, inherent in that system are *differences* as well as similarities in the properties and reactions of the members of any homologous series. A chemist, and it is from the standpoint of a chemist skilled in this art that the question of obviousness must be resolved, would consider the differences as well as the similarities in the properties and reactions of the members in any given homologous series..."

The courts, both before and after the enactment of Section 103, have determined the obviousness and patentability of new chemical compounds by taking into consideration their biological or pharmacological properties. Patentability has not been determined on the basis of obviousness alone. In fact, in many of the cases patentability has been found in spite of close similarity of chemical structure (*38*). At the same time, the mere showing of an unobvious or unexpected beneficial property in a new product does not of itself establish patentability. The consideration of other factors may be required (*10*). "A mere difference in degree (*38*) is not the marked superiority which ordinarily will remove the unpatentability of adjacent homologues of old substances." "Whether novel chemical compounds are patentable over prior art isomers and homologues (*7, 15-17*) is a question to be determined in each case."

In order to minimize patent prosecution problems of the type described, one should submit, along with the compound claims, composition and use claims. A new use for a known compound is properly claimed as a process under Title 35 U.S.C. 100(b) (*43*):

"The term 'process' means process, art or method, and includes a new use of a known process, machine, manufacture, composition of matter, or material."

In a recent case (*38*), the claimed compounds had been rejected as obvious under 35 U.S.C. 103 in view of a single reference showing trimethyl side chains corresponding to the triethyl and tributyl side chains on the compounds of applicant's invention. However, whereas applicant's products exhibited pharmacological properties, the prior art compounds were inactive.

In holding the compounds patentable, the court viewed the case as presenting the question of patentability "of a *new* chemical compound having an inherent *unknown*, unobvious pharmacologically advantageous property." But if the "advantage is not disclosed in applicant's application," he is "not in a favorable position (*18, 30*) to urge it as a basis for allowance of claims."

The importance of the differences in properties in compounds "structurally rather similar" was recognized by the Court of Customs and

Patent Appeals in still another case (*39*), in which the Court held that "in determining whether the claimed compounds are *obvious* within the meaning of 35 U.S.C. 103, (we think) their properties may and should be considered, . . ." Referring to an earlier decision of a case involving a similar problem (*26*), the court said, ". . . There is no evidence *in the record* which would lead one skilled in this art to expect that the differences in molecular structure between riboflavin and appellant's compound would cause this difference in properties (*38*)." The holdings in these cases (*17, 29, 33*) establish that too much legal significance should not be given to the bare term "homologue." The Court must apply the statutory test of obviousness under 35 U.S.C. 103, and "homology" is "nothing more than a fact which must be considered with all other relevant facts" in applying this test.

Prior Art Disclosures and Combining References

Another type of rejection for obviousness is based upon a combination of references and disclosures in the prior art of compounds which might readily be converted to claimed compounds. In one instance (*3*), reversing the examiner and the board, the Court of Customs and Patent Appeals' position was that "such conversion would not be obvious in the absence of any suggestion in the prior art as to why it should be made. . . . The mere fact that it is *possible* to find two isolated disclosures which might be combined to produce a new compound does not necessarily render such production obvious unless the art also contains something to suggest the desirability of the proposed combination."

In still other instances, rejections for noninvention have been based on the theory that it was obvious to make the claimed compound. Refuting this type of rejection (*27*), Chief Judge Worley (*38*) reasoned:

". . . the allowance of the claims to the compounds was based on the fact that they possessed unique, and presumably unexpected properties. *Since there was nothing to indicate that the compounds, when made, would have these properties, it was not obvious to make the compounds.* In such a case the allowance of claims to the compounds must depend on the proposition that *it was unobvious to conceive the idea of producing them*, within the meaning of Title 35 U.S.C., Section 103."

An organic chemist knows that other groups may be attached to a parent structure. This is true of all parent chemical structures. "Though this would be obvious to him, it does not follow that all new *compounds* so produced would be obvious in the sense of the patent law (*26, 38*)." As in a combination, using old elements, the question is whether the alleged invention would have been obvious to one skilled in the art. Invention under this test is a question of fact (*32*).

While knowledge of and complete familiarity with the chemical subject matter involved is of the utmost importance in dealing successfully with the issues involved, "the problem of obviousness under Section 103 in determining the patentability of new and useful chemical compounds, or, as it is sometimes called, the problem of chemical obviousness, is not really a problem in chemistry or pharmacology or in any other related field of science such as biology, biochemistry, pharmacodynamics, or ecology. It is a problem of *patent law* (*38*)."

Standard of Invention Is Subjective

It has been said authoritatively "that the (patent) Act of 1952 meant to change the slow but steady drift of judicial decision that had been hostile to patents which made it possible, in 1945, for Mr. Justice Jackson (*23*) in dissent to speak of the 'strong passion in this court (the Supreme Court) for striking them (patents) down so that the only patent that is valid is one which this court has not been able to get its hands on.' "

"The standard (of invention) is always subjective, the creature of an imagination projected upon the future out of materials of the past (*45*)."

The view of a nontechnical court confronted with the problem of determining the validity of a patent has been expressed by Judge Learned Hand of the Second Circuit (*42*):

"It is not for us to decide what 'discoveries' shall 'promote the progress of science and the useful arts' sufficiently to grant any 'exclusive right' of inventors (U.S. Constitution, Article 1, Sec. 2). Nor may we approach the interpretation of Sec. 103 of the Title 35 with a predetermined bias.

"The test laid down is indeed misty enough. It directs us to surmise what was the range of ingenuity of a person 'having ordinary skill' in an 'art' with which we are totally unfamiliar; and we do not see how such a standard can be applied at all except by recourse to the earlier work in the art, and to the general history of the means available at the time. To judge on our own that this or that new assemblage of old factors was, or was not, 'obvious' is to substitute our ignorance for the acquaintance with the subject of those who were familiar with it."

A careful and objective analysis of recent decisions both in the review tribunals of the Patent Office and in the courts, and particularly in the Court of Customs and Patent Appeals, indicates that there is an ever-increasing awareness of the complex problems uniquely peculiar to the chemical and related arts. The judiciary before whom the majority of contested cases are brought on appeal have unusual sophistication and expertise in the highly complex technical areas embracing the broad spectrum of chemistry and the chemical sciences. There is encouraging evidence of growing recognition that the interim results of research may properly lay claim to patent protection. A realistic treatment of the utility question in chemical cases appears to be more the rule.

The question of obviousness in chemical and pharmaceutical applications will continue to be decided not by arbitrary and mechanically applied rules but on the basis of the particular facts of each case. The aims and goals of research and of the patent system appear to be approaching a close correspondence. Even closer correlation lies just ahead.

Literature Cited[1]

(1) Aerosol Research Co. v. Scovill Mfg. Co., D.Ct. N.D. Ill., 137 U.S.P.Q. 701, at p. 723.
(2) Aluminum Co. of America v. Sperry Products, Inc., 285 F.2d 911, 917, 127 U.S.P.Q. 594, 399, Ct. App. 6 (cert. denied, 368 U.S. 890, 131 U.S.P.Q. 498).
(3) Bergel, In re, 292 F.2d 955, 130 U.S.P.Q. 206 (C.C.P.A. 1961).
(4) Birmingham v. Randall, 171 F.2d 957, 80 U.S.P.Q. 371.
(5) Bremner, In re, 182 F.2d 216, 86 U.S.P.Q. 74 (C.C.P.A. 1950).
(6) Bros Inc. v. Browning Mfg. Co. (8th Cir.), 137 U.S.P.Q. 624.
(7) Coes, In re, Loring Jr., 173 F.2d 1012, 81 U.S.P.Q. 369.
(8) Dodson, In re, 130 U.S.P.Q. 224, 292 F.2d 943 (C.C.P.A. 1961).
(9) Druey and Schmidt, In re, 138 U.S.P.Q. 39 (C.C.P.A. 1963).
(10) Finley, In re, 174 F.2d 130, 81 U.S.P.Q. 383, 386 (C.C.P.A. 1949).
(11) Goodyear Co. v. Ray-O-Vac Co., 321 U.S. 275, 60 U.S.P.Q. 386.
(12) Great Atlantic and Pacific Tea Co. v. Supermarket Equipment Corp., 340 U.S. 147, 71 Sup. Ct. 127, 95 L. Ed., Sup. Ct. Reports, 162, 87 U.S.P.Q. 303 (1950).
(13) Harrison and Packman, Ex parte, P. O. Bd. App., 129 U.S.P.Q. 172.
(14) Hartop and Brandes, In re, 135 U.S.P.Q. 419 (C.C.P.A. 1963).
(15) Hass, In re, 141 F.2d 122, 60 U.S.P.Q. 544.
(16) Henkel, Ex parte, 130 U.S.P.Q. 474.
(17) Henze, In re, 181 F.2d 196, 85 U.S.P.Q. 261.
(18) Herr, In re, 134 U.S.P.Q. 176 (C.C.P.A.).
(19) Hotchkiss v. Greenwood, 52 U.S. 248.
(20) Huyl & Patterson v. McDowell Co., Inc., CA 4, 137 U.S.P.Q. 620.
(21) Isenstead v. Watson, D.Ct. D.C., 157 F. Supp. 7, 115 U.S.P.Q. 408.
(22) Jaffee and Ogden v. Kassley, 137 U.S.P.Q. 653.
(23) Jungersen v. Ostby & Barton Co., 335 U.S. 572, 80 U.S.P.Q. 32, 36.
(24) Krementz v. S. Cottle Co., 148 U.S. 556.
(25) Krimmel, In re, 130 U.S.P.Q. 215, 292 F.2d 948 (C.C.P.A. 1961).
(26) Lambooy, In re, 300 F.2d 950, 133 U.S.P.Q. 270 (C.C.P.A. 1962).
(27) Larsen, In re, 292 F.2d 531, 130 U.S.P.Q. 209 (cert. denied, 133 U.S.P.Q. 209; cert. denied, 133 U.S.P.Q. 703).
(28) Levy, 30 J.P.O.S. 592 (1948).
(29) Lohr and Spurlin, In re, 137 U.S.P.Q. 548 (C.C.P.A. 1963).
(30) Lundberg, In re, 117 U.S.P.Q. 190.
(31) Mahn v. Harwood, 112 U.S.C. 354, 358 (1884).
(32) Maytag Co. v. Murray Corp. of America, Ct. App. 6, 137 U.S.P.Q. 819.
(33) Mills, In re, 126 U.S.P.Q. 513.
(34) Nelson, In re, 280 F.2d 172, 126 U.S.P.Q. 242 (C.C.P.A. 1960).
(35) Nichols v. Atkinson, 88 F.2d 688, 33 U.S.P.Q. 82.
(36) Novak and Hogue, In re, 134 U.S.P.Q. 335 (C.C.P.A.).
(37) Palmquist and Erwin, In re, 138 U.S.P.Q. 234 (C.C.P.A. 1963).
(38) Papesch, In re, 137 U.S.P.Q. 43 (C.C.P.A. 1963).
(39) Petering and Fall, In re, 301 F.2d 676, 133 U.S.P.Q. 275.
(40) Pieroh and Werres, In re, 138 U.S.P.Q. 238, 240 (C.C.P.A. 1963).
(41) Prusak, 35 J.P.O.S. 616 (1953).
(42) Reiner v. The I. Leon Co., Inc., 128 U.S.P.Q. 25 (Ct. App. 2).
(43) Riden and Flavin, In re, 138 U.S.P.Q. 112 (C.C.P.A. 1963).
(44) Rothermel and Waddell, Jr., In re, 276 F.2d 393, 125 U.S.P.Q. 328.
(45) Schaefer, Inc. v. Mohawk Cabinet Co., Inc., Ct. App. 2, 125 U.S.P.Q. 318, 320.
(46) Szwarc, In re, 138 U.S.P.Q. 208 (C.C.P.A. 1963).
(47) Twentier's Research Inc. v. Hollister Inc., Ct. App. 9, 138 U.S.P.Q. 473.
(48) Wilke and Pfohl, In re, 314 F.2d 558, 136 U.S.P.Q. 435.

---

[1] Legal abbreviations are defined on page viii.

RECEIVED October 31, 1963.

3

Inventorship in Chemical Patents

EDMUND A. GODULA

Parker & Carter, 8 So. Michigan Ave., Chicago, Ill.

> **Fact situations which make a person a sole inventor and which make two or more persons joint inventors, in the eyes of the law, are discussed. The concept of joint inventorship is contrasted with situations where others are "extended technical arms of the inventor." Problems such as the synthesis of a new compound by a chemist and discovery of a utility by a person trained in another discipline are presented, including selection of an old compound by a chemist for testing or a particular utility and discovery of a related utility by another person. Observations are made on the likelihood of a patent being held invalid under different improper inventorship situations.**

Imagine, if you will, a chemist telling a patent solicitor about new chemical compounds that were synthesized in his laboratory. He tells the patent solicitor that some of them have shown strong promise as plasticizers. He also says that some difficult problems had to be overcome in successfully synthesizing the compounds. He answers that after the first compound about 10 congeners were made and that the series as a whole is so promising that he has urged management to put several other chemists on the project.

The foregoing situation calls for blunt and pointed questions by the solicitor to find out:

Who conceived the first compound?
Who made the congeners?
Who solved the problems, if any, in synthesizing?
Were the congeners prepared by basically the same method used in synthesizing the first compound?
Were the compounds tested as plasticizers?
Did someone independently discover this utility?

Will the new variations be synthesized by all the workers following their own ideas or pursuant to a general plan?

There are other questions which a solicitor should ask an inventor or inventors, but the foregoing are presented merely to find out if there is one inventor or more for one invention or more.

Determining who is the inventor is an obligation imposed by the patent law, which is written generally under Title 35 of the United States Code. Section 115 of that title says that the applicant shall make oath that he believes himself to be the original and first inventor. Section 116 of the title says that when an invention is made by two or more persons jointly, each must sign the application and make the required oath. Every time a person is named as an applicant who is not the inventor or has not been named but is the inventor, there is potential trouble for the well being of any patent which issues. Courts have repeatedly invalidated patents for such reasons, and as recently as April 1962 a court in Massachusetts said that if a patent issues on an application of one who is not the true inventor, the patent is unauthorized by law and is void. It confers no rights on anyone and there can be no legal or equitable ownership of it (22).

To illustrate again how seriously courts can take the subject of inventorship, there was a case decided in 1936 (23) in which inventors called Fink and Udy claimed the same subject matter. Fink got a patent and Udy copied his claims. The United Chromium Co. owned the Fink application, and they bought the Udy application during the interference. According to the rules of the Patent Office, United Chromium was compelled to elect one of the applications, since it owned them both. They elected Fink's application, and the patent issued to him. Later on, a court held that Udy was the first inventor and not Fink. The court held the patent invalid.

Inventors' Extended Technical Arms

The question of who is the inventor alone and who are the inventors together is many times troublesome. Let us consider what courts have said about an applicant reasonably asserting that he is the sole inventor. First of all, courts have regularly said that a sole applicant must have a complete conception of the invention. A frequently quoted definition of conception of invention was stated in 1897 in a decision by the Patent Office Commissioner (*14*):

"The conception of the invention consists in the complete performance of the mental part of the inventive act. All that remains to be accomplished in order to effect the act or instrument belongs to the department of construction not invention. It is, therefore, the formation in the mind of the inventor of a definite and permanent idea of the complete and operative invention as it is thereafter to be applied in practice that constitutes an available conception within the meaning of the patent law."

This language says that the inventor must first of all have a conception or idea of a complete and operative invention. It does not mean that the invention, as it finally is made, has to conform in every respect with the conception. It may be modified to put it in better commercial form. The court which directly controls the Patent Office is called the Court of Customs and Patent Appeals (C.C.P.A.). This court has repeatedly referred to the quoted conception test. This court, however, has said that the language is modified (20) in that:

". . . the final size and shape of every part and location of every nut, screw and bolt may not be exactly forseen before conception of apparatus can be said to be complete. It is sufficient if the inventor is able to make a disclosure which would enable a person of ordinary skill in the art to construct the apparatus without extensive research or experimentation . . ."

Now this court was obviously talking about mechanical inventions since it referred to every nut and screw. But the same reasoning applies to chemical inventions. The court could almost be saying that:

"The conception of a new paint composition, having additives to make the paint more durable, does not have to detail the final concentration of the resin and the particular identity of all the solvents if a paint chemist of ordinary skill in the art is able to formulate the composition without extensive research or experimentation."

A case involved a high pressure syringe for injecting medicaments into a human. The sole inventor in the patent admitted he had a brainstorming session with G and the result was a contained chemical reaction system to propel medicament from a syringe. The inventor used G and others to help work out different embodiments. One assistant was alleged to have devised a propelling force through a chemical reaction which released carbon dioxide. Another assistant was to package the reaction in a cartridge for the syringe. The patent was attacked as not being a sole invention but a joint invention. The inventor was allowed to explain away the "brainstorming" session. The court was convinced that the inventor of the patent was indeed a sole inventor and further said that he was entitled to incorporate in the specification the suggestions of others who were working with him (3). The others were helpers—the extended technical arms. The court was influenced by the fact that the inventor had a reputation as an inventor, and the court was impressed by his established genius. On several occasions some courts have said that if there is an argument between two persons as to who is the inventor, and one person is particularly skilled in the art and the other is not, the presumption is that the skilled man is the inventor (6).

If an inventor sets up an experiment or a research program which leads immediately to a discovery by another, there may still be a single invention and a single inventor. Even if the inventor does not have a good idea of what might happen. A case went to the Supreme Court which involved the flotation process for ore separation. This was done in the art by mixing oil with the ore so that the metal plus the oil would separate from the rest of the ore. The inventors told a laboratory assistant to reduce the oil in the mixture gradually and make observations. When the oil content was about 0.5% of the ore, a froth developed which was found to contain more metal than before. The court held that the laboratory assistant was neither a sole inventor nor a joint inventor because the inventors who owned the patent planned the experiments which were in progress and directed the investigation day by day. These actual inventors conducted experiments largely by themselves and interpreted all the results. It just happened that the employee was placed in a position, so to speak, to make the crucial observation and discovery (*15*).

When the chemist acts as an extended technical arm or when he outdistances the reach of the arm is discussed in an 1868 Supreme Court decision (*1*), which is still applicable:

"Where a person has discovered an improved principle . . . and employed other persons to assist him in carrying out the principle, and they, in the course of experiments arising from that employment, make valuable discoveries ancillary to the plan of the employer, such suggested improvements are in general to be regarded as property of the party who discovered the original improved principle, and may be embodied in his patent. No suggestion from an employe not amounting to a new method or arrangement which in itself is a complete invention is sufficient to deprive the employer of the exclusive property of the perfected improvement."

Another old Supreme Court case (*17*) referred to the right of an inventor to use an extended technical arm. The court noted that to make an improved invention an inventor needs a considerable fund of knowledge, and where this fund is acquired before the invention is made those who imparted the fund to the inventor do not become joint inventors. During his experiments, the court said that an inventor seeks and secures a point from one scientist, another point from a machinist, another from a book, and so on. He is not any less an inventor for doing the first two things than he is for doing the last.

Another court (*8*) discounted a second chemist's contribution because it did not include the "whole essence" of the invention. The invention concerned a vanadium catalyst and carrier used in the process for making sulfuric acid. The first set of inventors came up with the idea of using an alkali in vanadic acid solution and fine carriers in which the particle size was no more than 60 microns. The particle size of the carrier was important. Another chemist told them to use kieselguhr as a carrier and potassium hydroxide as the alkali. The perfected invention, which was most useful, turned out to be made by combining potassium hydroxide with a vanadate salt on kieselguhr. The water evaporated, and the potash and the vanadium were present on the minute particles of kieselguhr throughout its mass. The court said that the idea of the chemist to use potassium hydroxide and kieselguhr did not include the "whole essence" of the invention. The evidence did not support that he was an inventor or a joint inventor.

Suggestion or Impetus Leading to Discovery

We are all familiar with the situation where chemist A brings a problem to chemist B or commissions B to undertake a project. A can introduce the project without giving any particular information by merely asking for an insoluble salt of compound X, or he can give a lead to a particular reactant. Is B an inventor or coinventor when he makes a successful compound?

Let us consider a case where one chemist, D, brought a problem to another chemist, B, saying he wanted to use a lead compound to make visible fingerprints, and he wanted to develop the same by use of a sulfite compound. B experimented for three years and finally originated and developed an invention for making fingerprints by first using a sensitizing solution with oleate of lead and then treating it with a developing solution containing soluble sodium sulfhydrate. The patent issued to D, the chemist who brought the problem and authorized the project.

The patent was held invalid because a wrong inventor was named. The court noted (*11*) that B worked a long time and rejected a lot of compositions before he developed the particular composition. Now, we do not know all the problems encountered, and we do not know whether the sulfhydrate and oleate of lead were critical; but this case certainly illustrates the problem and shows how a court may view the work of a chemist to whom a problem is brought as constituting something more than "an ancillary discovery."

Another case (*9*) considered a composite package of bugs to fix nitrogen for leguminous plants. Some bug species were good for one plant but not others. Different bugs were never used together because of inhibition of their nitrogen-fixing properties. A presented the general idea that it would be desirable to make a composite package to fix nitrogen on various leguminous plants. B devised such a package, and he was held to be the sole inventor because he did the experiments which led to the successful package of noninhibiting strains.

Another interesting nonchemical case (*18*) involved a person who came to a company and said he wanted them to design a six-wheel truck for him. Apparently, six-wheel trucks were not known at that time. The design turned out to be objectionable because it had a rigid dual axle.

Someone at the company suggested placing a universal joint at the axle, but this was rejected by the person who presented the problem and was paying for the work. This person then lost entire interest in the six-wheel idea and gave up. The company applied for and got a patent on a six-wheel truck with universal joint. This naturally led to a law suit and the court later held that the person at the company was the sole inventor because he was the originator of an improvement on a rejected idea, and he did not collaborate in its development with the person who originally brought the idea. This case introduced the idea of somebody presenting an idea of a less than complete conception and then abandoning it. The abandonment of invention usually has serious consequences in the patent law to the one who does the abandoning.

Joint Invention

Whether an invention is by a sole inventor or by a joint inventor, it is a mental product. It is either a sole or a joint mental product. Joint inventors are one legal animal, and a sole inventor is another legal animal. If chemist A is the applicant on one invention and chemists A and B are the applicants on another related invention, they are separate legal entities. To demonstrate dramatically how different, it has been held (7) that a joint invention of both A and B is a good anticipation to a later related invention of either A or B, and the fact that either A or B was a joint inventor of the reference does not discredit the reference in the slightest.

Generally, joint invention must involve some sort of collaboration between two or more inventors which leads to the product. Let us consider a textbook definition first. A very old and very often cited authority (*19*) says:

"Where two or more persons, acting jointly, conceive the same idea of means, they are . . . jointly entitled to the patent. The sphere of their joint labors and success is thus the mental part of the inventive act. That one conceives the idea and another reduces it to practice; that one conceives the principal idea and the other an idea which is ancillary to and inseparable from it; that one conceives one idea and the other a different idea, both of which are united in the concrete invention, neither of these are joint invention, nor do they give to the inventors the right to become joint patentees. Only where the same single, unitary idea of means is the product of two or more minds . . . is the conception truly joint . . .

". . . joint inventions are created . . . [either] . . . by complete mental development of the idea followed by reduction to practice, [or] . . . by the simultaneous operation of the physical and mental faculties, . . . the embodiment and conception advance side by side, and the completeness of the one is known to the inventors only from the successful practical application of the other. The concurrence of the inventors in the physical experiments by which the inventive act proceeds is essential to render the result a joint invention."

What does all this have to do with the problem of chemist B, who made the insoluble salt compound, when A's name is the only one on the patent application. There is no measure nor scale nor rational basis by which we can urge that an inventor must contribute at least 26.78% of the total product in order to earn a place as a joint inventor. The textbooks say, in different ways, that there must be true collaboration. Obviously, the joint inventors do not have to come onto the idea simultaneously. They do not have to work in each other's presence all the time, and they do not have to consult each other on all points. There is no decision which spells out fixed rules on how inventors become joint inventors, much less how chemists become joint inventors. There are decisions, however, which provide some guideposts.

The C.C.P.A. considered a case (5) in which neither of two joint inventors could pinpoint his exact contribution to the invention. It involved using soap as a catalyst under certain reaction conditions. There was a series of mutual discussions which led to the idea but neither could exactly state who came up with the crux. They admitted they probably did not contribute the idea of using soap in chorus, but they did insist that the inventive idea arose through discussion. The court repeated the old rule that there is a heavy presumption that when an application says the inventors are joint it is believed. It is a big burden to try to up set this presumtion. Since there is such a presumption, it is safer to name joint inventors when in doubt. Courts rather consistently take a more strict view of a patent which has left off a true joint inventor rather than one which includes someone who was not an inventor. A district court (21) has said that a person is a joint inventor even if his contribution is minor, as long as it was required to complete the invention or make it operable.

Chemical Problems and Joint Inventions

Chemists, like other inventors, may come to an honest difference of opinion as to which of, say, two chemists made the more important contribution to an invention. It is generally true that each believes his contribution to be the more important. Suppose one chemist discovers a new process for making an old compound in which less expensive starting reactions could be used. Say, the yield is insignificant to a degree which makes the process practically useless. Chemist B finds that by controlling the pH and concentrations within a given range he can increase the yield to 88%. Each will likely insist that his contribution is the more important but, in any event, this is a case of joint invention. If each chemist went his own way and filed his own patent application, there would be an interference, and the Patent Office Board of Interferences has said, under such general situations (16), that the issue is whether the second chemist's contribution, after the first showed him the process, involved only the skill of the ordinary man in this art or such a material contribution as to make him a co-inventor. Chemist *A* and chemist *B* working together have made a joint invention. One would be no place without the other, practically speaking. This is similar to the six-wheel truck case, except that here there is no feature of abandonment.

A similar case would be where chemist *A* conceived a structural formula for a new compound but had trouble making it by the conventional chemical processes which first occurred to those skilled in the art. Chemist *B* following some nonobvious process steps does make the compound. Again one would be no place without the other, and this is a proper joint invention. A court has considered an analogous type of situation (*26*) and said that when one can perceive the crude form of elements or possibility of adaption to accomplish a result, he becomes a joint inventor with the one who actually does accomplish it.

Suppose we have a problem in which pharmaceutical chemists try to stabilize a blood anticoagulant in a pharmaceutical composition. Suppose two pharmaceutical chemists are assigned to solve the problem, and suppose one does get what could be called the "best thought" that went into the solution. Let us say the use of a hyposulfite as a stabilizing ingredient which, unfortunately, clouded up the pharmaceutical composition. This would not affect the usefulness but would be objectionable to doctors, therefore, unsalable. The second chemist found a solution to the clouding up.

This type of situation has been recognized as a joint invention because a court has said (*24*) that the mere fact that one of two joint inventors conceived the "best thought" that went into the invention does not invalidate the patent. Both thoughts make up a joint invention, even where a series of steps are present in a process or a number of elements in combination.

Let us take another hypothetical case. An often recurring problem occurs with a discovered prototype compound which is active, say, as a hypotensive agent. It is necessary to investigate the possible variations which may have equivalent or better activity. Three or four chemists are put on the job of investigating variations. They work together every day for a long period and discuss matters together, confer, and exchange ideas. One suggests and does one thing, another does another thing, one supplementing the work of another, and so on. Say they use a common notebook or several notebooks with cross references to each other. This general type of activity was held to support a joint invention in a case involving photographic sound reproduction or talking movies (*2*). Of course it is hazardous to use this as a general rule and make joint inventions on all types of analogs and homologs and isomers which come out of a particular project. When an assistant down the line merely extends an alkyl side chain from methyl to propyl on the general directions of a first assistant, then joint invention is questionable.

Chemical Utility and Joint Inventions

It is legal truism that you do not have a chemical invention until you show that the invention is useful for something—that is, it has utility as required by the patent laws. Consider the situation where a chemist synthesizes a novel structure and has absolutely no idea of a use for it. This same chemist puts it through a series of tests which he prescribes, and a utility is uncovered. The evidence is excellent that he is a sole inventor. If such compounds are run through standard test procedure, then these standard testing procedures may be considered as extended technical arms of the chemist. He is still the sole inventor. But if the standard testing procedures fail to uncover utility, there is an uncompleted invention. If someone else comes along and suggests a new test to evaluate another utility not in the standard testing procedure, and uncovers it, there is strong evidence for a joint invention.

A pertinent decision (10) regarding the discoverer of a utility involved an envelope or container for a tea bag. One inventor had made an envelope having unspun fibers thermoplastically bonded together so that they would be watertight. A problem arose when they were infused in hot water because toxic substances were released which affected the taste. The problem of making it tasteless remained, until another inventor solved it by choosing a proper plasticizer and properly bleaching the fibers. The court noted that the inventor must be one who produces a thing which is useful and that the second chemist by making it useful became one of the inventors. This case illustrates the point of utility being necessary as an ingredient, although the case was actually resolved on the proposition that the bag was still useful for infusions irrespective of taste properties.

Features of Inventorship

Occasionally courts speak about a single-idea invention, and there is a presumption that with such single-idea inventions, joint inventorship is unlikely. An example would be a known biological method for producing an antibiotic. The antibiotic yield is increased by, say, merely raising the temperature. It would be difficult to argue that this is a joint invention.

A disadvantage of joint invention which commonly arises is the failure of corroboration in a later interference. Many times joint inventors are the only ones who have observed the work which led to the invention, but these people cannot be used to corroborate the fact of invention to show priority because they are interested parties.

Another difficulty with joint inventors arises when it becomes necessary to file a continuation case. This may be embarrassing because of the requirement to keep the same inventor entities in order to get the benefit of their earlier filing date. Of course the disadvantages and objections should be disregarded if the situation requires that joint inventors be named. If a situation is questionable, it would be wiser to name joint inventors where the actual or essentially complete contribution of a single inventor would be hard to establish.

The hazards of misjoinder or nonjoinder of inventors are reduced because the 1952 Statute permits addition or removal of inventors upon a showing that a misjoinder or a nonjoinder occurred without any deceptive intention. This may be done while the application is pending (35 U.S.C. 116) or even after patent issues (35 U.S.C. 256). The patent solicitor is therefore given a chance to correct any innocent mistake. The mistake must be innocent, or else a conversion will not be permitted. A general manager removed the name of one joint inventor at time of filing and then tried to add his name before suing an infringer. A court prevented the conversion (*4*) saying it was an error of judgment and not a mistake. If the evidence indicates that the misjoinder or nonjoinder was intentional with an intent to deceive, no recourse could be made to the corrective provision in the statute, and the validity of the patent could be later challenged. The C.C.P.A. (*25*) has said, however, that any attempt to convert the inventor entity should be timely, and that lack of diligence can prevent such attempted conversion.

Apparently you cannot become too casual about changing the inventor entity. A respected text writer (*13*) says that the Patent Office would not permit a new sole inventor to be named if he never appeared on the application as an inventor. He could be named as a joint inventor but not as a single inventor.

Since the inventor entity can be changed, any chemist learning of a filed application, say in his company, should bring any pertinent facts he may know to the attention of the patent solicitor at any time.

Another feature of joint inventions relates to the claims in the application. All the claims should ideally cover inventions which are truly joint. If any claim is directed to a feature which was contributed by one inventor alone, such claim may be held invalid (*12, 21*).

A final word about the order of names of joint inventors which appear on a granted patent. The sequence of names follows the order in which they appear in the application oath. In the eyes of the law, the first name on a patent does not imply in any way a greater contribution or lesser or what have you. All joint inventors are considered as jointly making the invention somewhat in the manner of a joint tenancy holding real estate. Each has an undivided part in the whole as well as in the part.

Unfortunately, glory goes to Jones and obscurity to *et al.* This is a truly unhappy occurrence which can be remedied partly by extra compensation for the inventors *et al.*

Conclusions

The selection of a proper inventor entity is still a serious consideration despite the liberality in changing inventors permitted by the statute. The validity of a patent is most in danger with a single inventor named as a wrong inventor or if any inventor entity, sole or joint, is deliberately misnamed for any reason. The courts in general take a more serious view of an omitted joint inventor rather than one added who is not in fact a joint inventor. There is heavy presumption that the inventors named in the patent are the correct ones. Many courts do not like to see a patent attacked because of improper inventorship and such courts regard this merely as a technical defense. In general, if there is any doubt, it is recommended that the patent solicitor select a joint inventor entity rather than a sole inventor entity.

The chemical art creates great problems in determining a correct chemical inventor entity. One built-in problem is the frequent later determination of a utility for a chemical compound or composition. An invention is not complete until its usefulness is determined, and this utility may be established either by an extended technical arm of the chemist who made the compound or by someone who is not such an arm.

Other problems arise with new chemical compounds and compositions on one hand, and methods to make these compounds and compositions on the other hand. Every chemist knows that the first devised process has a lot of bugs, and it is necessary to determine proper reaction conditions such as concentration, temperature, and the like. Whether or not determination of such conditions is the work of an extended technical arm presents other problems. The determination of the chemical inventor entity is often a difficult problem, and it remains difficult even when all the facts of a particular situation are presented to a solicitor. Most often, the chemist inventor, the chemist joint inventors, and the chemist's technical arm will themselves appreciate what is the correct inventive picture. Chemists have an interest in this problem and should not hesitate to plead their case to the patent solicitor. Solicitors should not be overly officious in pronouncing an inventor entity but should be liberal and have an eye turned towards the morale, working relationships, and working conditions of the chemist. Liberality of course should never become laxity, which would endanger any issued patent.

Literature Cited[1]

(1) Agawam Co. v. Jordan, 7 Wall 583 (1868).
(2) Altoona Publixs Theaters Inc. v. Tri-Ergon (3rd Cir.) 22 U.S.P.Q. 8.
(3) Becton-Dickinson & Co. v. R. T. Scherer Corp., (D.C. Mich.) 94 U.S.P.Q. 138.
(4) Blue, John Co. v. Dempster Mill Mfg. Co. (D.Ct. Nebr.) 172 F. Supp. 23.
(5) Brown v. Edeler, 110 F.2d 858, C.C.P.A.
(6) Davis v. Carrier, 28 U.S.P.Q. 227, C.C.P.A.
(7) Dwight & Lloyds Sintering Co. v. Greennalt (2d Cir.) 27 F.2d 823.

(8) General Chemical Co. v. Standard Wholesale Phosphate & Acid Works (D.Ct. Md.) 36 U.S.P.Q. 472.
(9) Kalo Inoculant Co. v. Funk Bros. Seed Co. (7th Cir.) 161 F.2d 981.
(10) Kendall Co. v. Tetley Tea Co. (D.Ct. Mass.) 85 U.S.P.Q. 298.
(11) Kuhne Identifying Systems Inc. v. United States (Ct. Cl.) 28 U.S.P.Q. 151.
(12) Larsen Products Corp. v. Perfect Paint Products, Inc. (D.C. Md.) 191 F. Supp. 303.
(13) McCrady, "Patent Office Practice," 4th ed., Typecraft, Inc., 1959.
(14) Mergenthaler v. Scudder, C.D. 724 (1897).
(15) Mineral Separation v. Hyde, 242 U.S. 261.
(16) Nielsen v. Cahill, P.O. Bd. Interfer., 133 U.S.P.Q. 563.
(17) O'Reilly v. Morse, 14 L. Ed., 601 (1853).
(18) Pointer v. Six Wheel Corp., 83 U.S.P.Q. 43.
(19) Robinson, "Patents," Vol. 1, 567, Little, Brown & Co., 1890.
(20) Tansel, In re, (C.C.P.A.) 117 U.S.P.Q. 188.
(21) Thropp & Son v. DeLaski and Thropp Circular Woven Tire Co., 226 Fed. 941.
(22) Tracer Lab. Inc. v. Industrial Nucleonics Corp. (D.Ct. Mass.) 133 U.S.P.Q. 306.
(23) United Chromium v. General Motors Corp. (2nd Cir.), 31 U.S.P.Q. 105.
(24) United Shirt & Collar Co. v. Beattie (2nd Cir), 149 Fed. 736.
(25) Van Otteren v. Hafner, C.C.P.A. 278 F.2d 738.
(26) Vrooman v. Penhallow, 179 Fed. 296.

---

[1] Legal abbreviations are defined on page viii.

RECEIVED October 31, 1963.

4

Ownership in Inventions

J. R. JANES

Watson, Leavenworth, Kelton & Taggart[1],
100 Park Ave., New York, N. Y.

> To the general rule regarding ownership of inventions, there are two important qualifications. First, if the inventor creates his invention under circumstances such that another supplies material, money, or other aid, that person, party, or company may become entitled to a personal shop right to use the invention free of liability to the inventor. This shop right is not assignable as such to others, nor can it be licensed, but it can be transferred in a sale of the whole assets and business of the holder. Second, if the inventor makes the invention under circumstances such that he has sold his services in inventing to another party or person, that person or party may become the owner of the full title to invention.

There can be many factual situations in which an invention is made and in which a question of title can arise. Here, the author indicates the ones that are sharply delineated—in black and white—and then only briefly looks into the grey areas that so complicate problems of ownership.

An Invention Is Property

Inventions are property and are generally owned by the person who created them; they receive protection under the law in the same way that other property receives protection. However, the property in an invention that is the subject of protection under the law is not the idea behind the invention but the tangible reduction to practice

---

[1] Present address: Janes & Aeschlimann, 70 Pine St., New York, N. Y.

of the idea. Conception alone is not protected. It is in fact no more than an idea for an invention to be made but not completed. The law does not protect the mere conception of an invention, but it does protect the completed invention—that is, the work product of that conception. This means, of course, that at common law, after the idea or principle behind an invention is made known, someone else is free to use that idea or principle to make another invention or even possibly another embodiment of the first invention.

Thus, under common law, prior to patent statutes, an inventor owned the invention he created, but he could protect it as a practical matter only to the extent that he could keep it secret. For instance, in the case of a chemical formulation, an inventor could protect his new formulation only as long as he could keep the public from ascertaining what the formulation was. If it was susceptible to having its composition determined by analysis, then after the formulation was put on the market, he took the risk of letting it become known so that others could duplicate it. On the other hand, if he could keep it secret, then he could retain his property right more or less indefinitely. Two outstanding examples of secret chemical formulations are Coca-Cola and Smith Brothers cough drops. It is still the law today that an inventor can protect his formulations by keeping them secret.

Patents Give Special Protection

Obviously, keeping inventions secret is not in the public interest. If the public can learn what is the nature of the invention, it is benefited greatly, because others can then build on this knowledge to make further improvements in the art. Furthermore, the necessity of keeping an invention secret does not afford worthwhile protection for the invention, for there is always the risk that the secret will be lost. ·Thus, many countries many years ago adopted patent laws which give to an inventor who chooses to take advantage of them a monopoly in a new invention for a limited time in exchange for full disclosure of the invention in a patent and a dedication to the public of the free right to make, use, and sell that invention after the patent has expired. Thus, in return for a disclosure of the invention, the patent laws give to the inventor the right to maintain exclusivity for the term of the patent even though it is no longer secret, and as a result an inventor now has the choice either of keeping his invention secret or of taking advantage of the patent statute.

A Federal court (3) has summarized the parallel protection afforded by trade secrets and patents as follows:

"However different these concepts of trade secrets and patents may appear to be, there is an important similarity; they are both means to competitive advantage. The value in both lies in the rights they give to their owners for monopolistic exploitation. The owner of a patent can make something which no one else can make because no one else is permitted. But circumstances are frequently such that the owner of a trade secret can make something which no one else can make because no one else knows how. The patent owner has a monopoly created by law; the trade secret owner has a monopoly in fact. In both cases there exists the possibility of either limited or complete transfers of the right to the exclusive use of an idea."

First U.S. Patent Law Enacted in 1790

The first U.S. patent law was enacted on April 10, 1790, and there have been many patent laws since. The latest is the Patent Act of 1952. All of them implement the constitutional power to grant patents to the inventors of the discoveries which are the subject of their patents.

Both the Constitution and the first Patent Act contemplated the simple situation of an inventor making an invention himself, reducing it to practice, and then proceeding with the filing of an application for a patent, which would have been prepared, at first, by himself and later on by patent agents or attorneys.

This simple picture of an inventor patenting his own inventions was to change rather quickly with the development of the industrial revolution. An inventor who has made a valuable discovery and obtained a patent usually wants to exploit his invention, and to do this he forms a business. Perhaps his discovery was a process of vulcanization of rubber. As a result of that, he founded a company to vulcanize rubber. If the company was successful, it and its line of products grew, and as it did the number of employees increased. Rather quickly, situations rose wherein inventions were made by employees of these new companies. An employee might be in charge of the operation of one of the pieces of chemical equipment of the plant, for example, but in the course of this operation, he might think of a way of improving the process. As a result of these changes in the economic picture, the courts were forced to develop equitable rules determining the ownership of inventions made by employees of corporations. These were designed to fit the various types of circumstances that arose.

Shop Right

One of the easiest types of situations to resolve was the case where the employee was an ordinary employee engaged in carrying out general duties within the company, who made an invention in the course of his employment on company time and at company expense. The shop-right rule in such a case is well set out by the Supreme Court in United States v. Dubilier Condenser Corp. (8):

". . . where a servant, during his hours of employment, working with his master's materials and appliances, conceives and perfects an invention for which he obtains a patent, he must accord his master a non-exclusive right to practice the invention. *McClurg v. Kingsland*, 1 How. 202; *Solomons v. United States*, 137 U.S. 342; *Lane & Bodley Co. v. Locke*, 150 U.S. 193. This is an application of equitable principles. Since the servant uses his master's time, facilities and materials to attain a concrete result, the latter is in equity entitled to use that which embodies his own property and to duplicate it as often as he may find occasion to employ similar appliances in his business. But the employer in such a case has no equity to demand a conveyance of the invention, which is the original conception of the employee alone, in which the employer had no part. This remains the property of him who conceived it, together with the right conferred by the patent, to exclude all others than the employer from the accruing benefits. These principles are settled as respects private employment."

This is a simple definition of what is now known as shop right and is also a summary of the shop-right rule which had been applied by a large number of courts prior to this case.

In the Dubilier case, the two inventors, Dunmore and Lowell, were employed by the National Bureau of Standards. The bureau at that time was composed of a number of divisions, one of which was the electrical division, which was further subdivided into sections, of which one was the radio section.

Dunmore and Lowell were employed in the radio section, and engaged in research and testing in the laboratory. In the outlines of laboratory work, the subject of "airplane radio" was assigned to the group of which Dunmore was chief and Lowell a member. The subject of "radio receiving sets" was assigned to another group to which neither Lowell nor Dunmore belonged.

In the summer of 1921, Dunmore, as chief of the group to which "airplane radio" problems had been assigned, without further instructions from his superiors, picked out for himself a problem assigned to the Bureau by the Navy—that of operating a relay for remote control of aerial bombs and torpedoes—"as one of particular interest and having perhaps a rather easy solution, and worked on it." In September he solved it.

In the midst of these aircraft investigations and numerous routine problems of the section, Dunmore was wrestling in his own mind, impelled thereto solely by his own scientific curiosity, with the subject of substituting house-lighting alternating current for direct battery current to power radio apparatus. The conception of the application of alternating current concerned particularly broadcast reception and was in no way related to the remote control relay devised for aircraft use. This idea was conceived by Dunmore Aug. 3, 1921, and he reduced the invention to practice Dec. 16, 1921. Early in 1922 he advised his superior of his invention and spent additional time in perfecting the details. On Feb. 27, 1922, he filed an application for a patent.

While performing their regular tasks, Dunmore, together with Lowell, experimented at the laboratory in devising apparatus for operating a radio receiving set by alternating current with the hum incident thereto eliminated. The invention was completed on Dec. 10, 1921. Before its completion no instructions were received from and no conversations relative to the invention were held by these employees with the head of the radio section or with any superior.

They also conceived the idea of energizing a dynamic type of loud speaker from an alternating current house-lighting circuit, and reduced the invention to practice on Jan. 25, 1922. On March 21, 1922, they filed an application for a "power amplifier." The conception embodied in this patent was devised by the patentees without suggestion, instruction, or assignment from any superior.

Dunmore and Lowell were permitted by their chief, after the discoveries had been brought to his attention, to pursue their work in the laboratory and to perfect the devices embodying their inventions. No one advised them prior to the filing of applications for patents that they would be expected to assign the patents to the United States or to grant the Government exclusive rights thereunder.

The court held on these facts that the inventors Dunmore and Lowell owned the inventions which they had made and that the only right the Government had was a shop right—that is, the free right to use the invention which Dunmore and Lowell had made.

Shop Right Denied

In Heywood-Wakefield Co. v. Small (5), the Court of Appeals of the First Circuit held that the employer was not even entitled to a shop right in the case where the employee had completely conceived of and developed the invention at home on his own time. The inventor, in that case Small, worked in the plant of Heywood-Wakefield, where it manufactured car seats for trolley cars and railroad coaches. Small was a checker, whose duties were to check every part of the car seat and all parts of the car seat wherever they were made and to see that the goods were made according to the standards of the company. He was never assigned any work on improving the company's product, but he knew that the car-seat base then manufactured by the company was unsatisfactory. Consequently, he worked on a method of improving this base, and he did this work at home after the idea for the resolution of the problem had been suggested to him by the mechanism of an electrical toaster. Finally, in his home workshop, he evolved what he believed to be a satisfactory reversible car-seat base, and he made a blueprint and a cardboard model of his invention, which he brought in to show to his superior. Further development was then carried forward by the inventor in cooperation with the company.

The court held that under these circumstances, the invention and the patent which he obtained for it were certainly the property of Small, and that the company was not even entitled to a shop right. The court found that the invention had been completed by the inventor before he had advised the company of its existence and that further work at company expense on it was only for the benefit of the company in developing it to a state where it could be put in commercial production. This was a close case, because there was a dissenting opinion by one of the three judges on the court who thought that the employer was entitled to a shop right on the basis that a considerable amount of money had been put into the development of the invention after the inventor had advised the company of it. It was in a very crude form at that time.

In Bowers v. Woodman (*1*), the inventor in question had originally been employed by the company, not because of his technical skill but as an ordinary employee. When he entered the employ of the company, it was not even known that he possessed any aptitude as an inventor. He was promoted to the position of superintendent on the basis of his ability, and he then assumed the duties of generally supervising the operations of a manufacturing plant. The Court pointed out:

"The respondent was not originally employed because of his technical skill. When he entered the employ of the Wickwire Spencer Steel Company, it was not known that he possessed any aptitude as an inventor. He was promoted to the position of superintendent, and he assumed the duties of generally supervising the operations of a manufacturing plant. Quite naturally, it was his duty, among others, to see that the products of the plant kept pace with the demands of the trade; that research and experiments were conducted with a view to discovering and developing improvements upon the product. If, in the performance of these duties, he developed a talent for working out novel and useful devices, it does not follow that he was employed to invent any specific device. If he had never invented anything, he could not have been charged with a failure in the performance of his duties as superintendent, or with a failure to fully earn his compensation. Not only is there no contract to assign his inventions, but there is, in this case, no contract to invent. In the absence of either of such contracts, the great weight of authority is to the effect that the employer has an irrevocable license to use the invention, but has no rights to compel a conveyance of the patent covering the invention."

Ownership of Inventions Made by Employees Hired to Invent

The Bowers' decision approaches a new problem which arose in the relations between employer and employee as a result of a further refinement and specialization of employee tasks. With the development of the modern corporation, there arose also a new class of employees, those hired for the particular purpose of inventing or improving existing processes or products of the company. In developing the principle of a shop right, the courts had recognized that the employer, because of his financial contribution to the creation and development of the invention, was entitled to something for his investment. They did not feel it necessary or desirable to transfer ownership of the invention to the employer because of the constitutional provision that the inventor should receive the patent and also because of the peculiar nature of invention. In the words of the Supreme Court, in the Dubilier case (*8*):

"The reluctance of courts to imply or infer an agreement by the employee to assign his patent is due to a recognition of the peculiar nature of the act of invention, which consists neither in finding out the laws of nature, nor in fruitful research as to the operation of natural laws, but in discovering how those laws may be utilized or applied for some beneficial purpose, by a process, a device or a machine. It is the result of an inventive act, the birth of an idea and its reduction to practice; the product of original thought; a concept demonstrated to be true by practical application or embodiment in tangible form. *Clark Thread Co. v. Willimantic Linen Co.*, 140 U.S. 481, 489; *Symington Co. v. National Castings Co.*, 250 U.S. 383, 386; *Pyrene Mfg. Co. v. Boyce*, 292 Fed. 480, 481.

"Though the mental concept is embodied or realized in a mechanism or a physical or chemical aggregate, the embodiment is not the invention and is not the subject of a patent. This distinction between the idea and its application in practice is the basis of the rule that employment merely to design or to construct or to devise methods of manufacture is not the same as employment to invent."

In the case of an employee hired to invent, however, the courts faced a new problem, because here the inventor had actually offered, as his services for which he was compensated by his salary, his originality of thought and his ability to develop new things. In effect, the employee had contracted to make inventions in return for his salary or other compensation. In these circumstances, the courts decided that the employer was entitled to title to the invention itself and to any patents derived from it. Again, in the words of the Supreme Court in the Dubilier case (*8*):

"One employed to make an invention, who succeeds, during his term of service, in accomplishing that task, is bound to assign to his employer any patent obtained. The reason is that he has only produced that which he was employed to invent. His invention is the precise subject of the contract of employment. A term of the agreement necessarily is that what he is paid to produce belongs to his paymaster. *Standard Parts Co. v. Peck*, 264 U.S. 52. On the other hand, if the employment be general, albeit cover a field of labor and effort in the performance of which the employee conceived the invention for which he obtained a patent, the contract is not so broadly construed as to require an assignment of the patent."

Chemist Hired to Invent

An example of the type of factual situation to which the court was referring in the Dubilier case is found in Houghton v. United States (6). Houghton was a trained chemist holding a degree from a university, and he was appointed assistant chemist in the office of Industrial Hygiene and Sanitation in the U. S. Public Health Service (USPHS). His duties consisted chiefly in analyzing samples of dust from industrial plants. Dr. Cumming, the U.S. Surgeon General under whom he was working, conceived the idea of combining an irritant gas with hydrocyanic acid gas so as to produce a safe fumigant. The use of cyanogen chloride gas as the irritant with the deadly gas had been suggested in a German periodical, and experiments and studies along that line were being conducted at the direction of USPHS. Houghton therefore was assigned the task of conducting experiments under the direction of the Surgeon General for the purpose of determining how best to produce and combine the gases to achieve the result which the Surgeon General had in mind. For this purpose, he was relieved of his work and sent to Edgewood Arsenal to make the experiment. His regular salary was paid to him while he was thus engaged. The court held:

"But the case here presented is that of an employee who makes a discovery or invention while employed to conduct experiments for the purpose of making it. Houghton did not conceive the idea of combining an irritant gas with hydrocyanic acid gas, so as to produce a safe fumigant. That was the idea of Dr. Cumming, the Surgeon General, under whom he was working. He did not conceive the idea of using cyanogen chloride gas as the irritant with the deadly gas. That idea had been advanced in a German periodical, and experiments and studies along that line had previously been conducted at the direction of the Health Service. All that he did was to take the idea of the Surgeon General, upon which the Health Service had been experimenting, and conduct experiments under its direction, for the purpose of determining how best to produce and combine the gases so as to achieve the result which the Surgeon General had in mind. For this he was relieved of other work and sent to the Edgewood Arsenal to make the experiments. His regular salary was paid to him while he was thus engaged, and, when he deduced from the experiments the method to be followed in producing and combining the gases, he did merely that which he was being paid his salary to do. Under such circumstances, we think there can be no doubt that his invention is the property of his employer, the United States, *U.S. v. Solomons,* supra; *Gill v. U.S.,* supra, 160 U.S. 426, 435, 436, 16 S.Ct. 322, 40 L.Ed. 480; *Standard Parts Co. v. Peck,* 264 U.S. 52, 44 S.Ct. 239, 68 L.Ed. 560, 32 A.L.R. 1033.

"The rule applicable in such cases cannot be better stated than it was by Mr. Justice Brewer in the *Solomons* case, supra, where he said [at page 346 (11 S.Ct. 89)]:

"'An employee, performing all the duties assigned to him in his department of service, may exercise his inventive faculties in any direction he chooses, with the assurance that whatever invention he may thus conceive and perfect is his individual property. There is no difference between the government and any other employer in this respect. But this general rule is subject to these limitations. If one is employed to devise or perfect an instrument, or a means for accomplishing a prescribed result, he cannot, after successfully accomplishing the work for which he was employed, plead title thereto as against his employer. That which he has been employed and paid to accomplish becomes, when accomplished, the property of his employer. Whatever rights as an individual he may have had in and to his inventive powers, and that which they are able to accomplish, he has sold in advance to his employer.'"

The Houghton situation is the normal situation in a chemical research laboratory.

In Dinwiddie v. St. Louis & O'Fallon Coal Co. (2), the Court of Appeals of the 4th Circuit found that consultants hired to develop a certain process for the coal company, who were paid a per diem salary and expenses and whose total expenditures in the development were borne by the coal company, were obligated to assign to the company any patents which were developed in the course of their research.

Employer's Assignment Agreement

The cited cases were quite clearly and easily decided on their facts. However, there can be all kinds of factual situations, ranging from the clear case of shop right to the clear case of being hired to invent, in which the courts would have great difficulty in deciding whether or not a company was entitled to the invention which the employee has made or whether it was only entitled to a shop right. This, of course, led to much uncertainty on the part both of employers and employees as to which of them was entitled to the title to the invention in doubtful cases. In this circumstance, the concept arose that the employee should agree in advance, as a condition of this employment, that he would assign to the company any inventions which he made in the course of his employment. Thus, the employee's invention assignment agreement came into use. Such agreements have been held by the courts to be valid and enforceable, provided their provisions do not effect such a restraint upon the employee or the employer as to be in contravention of public policy.

When an employee is hired to invent, all that the employee's assignment agreement really does is to bring emphatically to the employee's attention his obligation to assign the inventions that he makes. However, by a contract of this type the company can also ensure itself of the assignment of inventions made by employees in categories where they may not clearly have been obligated under the law to assign to the employer any inventions they might make. Thus the employees' invention agreement is intended to and does eliminate the grey area, where it is not clear whether the company is entitled to inventions made by employees not specifically or clearly hired to invent.

In addition to the patent clause requiring the assignment of inventions made in the future by the employee, an employee's assignment agreement normally also includes a secrecy clause which obligates the employee to keep secret the employer's know-how, trade secrets, or other confidential or secret information. This also is an obligation which the employee has under the common law, and the agreement is not really necessary to implement it, but it again makes the employee's obligation clear to the employee and also facilitates maintenance of its rights by the company in the case of an erring employee.

Examples of Assignments

The following are the patent clauses of a rather simple employee's assignment agreement:

"I will hold solely for your benefit and will fully and promptly disclose to you and assign in writing to you without additional payment all of my right, title and interest in and to all those discoveries, inventions and improvements which have been or shall be made, conceived or reduced to practice by me, either alone or with others, in the courts of my employment with you and which fall within the scope of your business activities, investigations, or research programs, as heretofore or hereafter conducted or definitely contemplated, and whether made within or outside of my usual work hours and whether on or off your premises.

"I will both during and after termination of my employment with you assist you in every proper manner, and at your expense and without cost to me, to obtain for you in any and all countries and to maintain and enforce patents on all the discoveries, inventions and improvements assigned by me to you as above provided."

The first paragraph obligates the employee to disclose and assign, without additional payment other than his salary, all of his right in the discoveries, inventions, and improvements which he may make in the course of his employment. This obligation is limited to inventions, discoveries, and improvements falling within the scope of the company's business activities, investigations, or research programs, and such an obligation has been held to be enforceable by the courts as reasonable. In the second paragraph, there is an obligation to assist the employer in obtaining, maintaining, and enforcing patents upon these inventions. This is essential because, as we have seen, the Constitution provides only for issuance of patents to the inventors.

The third paragraph is the secrecy clause:

"I will never use or divulge without your written permission any information, know-how, data or other knowledge not already available to the public respecting such discoveries, inventions and improvements or your business methods or systems, or your trade secrets, or confidential or secret information, or other private or confidential matters relating to your business activities, investigations or research programs, which may have become known to me or which I may have acquired during my employment with you for any reason whatsoever. I will not retain the possession of, or remove without the written consent of one of your executive officers, any reproduction or any record or copy of any such information, knowledge or data."

This obligates the employee never to use or divulge without the permission of the employer any of the employer's secret information. This type of clause is also regarded as reasonable and enforceable by the courts. This clause, of course, does not apply to information that is not kept secret, such as the information published in a patent or in the literature. Later publication of previously secret information could also serve as a release to the employee as to such information as of the date of publication.

Unenforceable Assignment

The following is an illustration of a type of patent clause in an assignment agreement which the Courts regard as unreasonable and therefore unenforceable. In Guth v. Minnesota Mining & Manufacturing Co. (4), the Court of Appeals, 7th Circuit, was faced with a contract which obligated the inventor to assign:

"(a) all my rights to inventions which I have made or conceived, *or may at any time hereafter make or conceive,* either solely or jointly with others, relating to abrasives, adhesives or related materials, *or to any business in which said company during the period of my employment by said company or by its predecessor or successor in business,* is or may be concerned, and

"(b) all my rights to inventions which, during the period of my employment by said company or by its predecessor or successors in business, I have made or conceived, *or may hereafter make or conceive,* either solely or jointly with others, or in the time or course of such employment, or with the use of said company's time, material or facilities, or relating to any subject matter with which my work for said company is or may be concerned; and . . ."

The italicized clause in question requires that the employee assign all inventions that he might make or conceive at any time after the date of the agreement—that is, even after his employment by the company had reached its end—as to any subject matter with which the company was or might be concerned. In other words, there was no time limit set on this obligation, nor was its scope limited. The court found this bad, holding as follows:

"Applying the rules of these decisions to the contract under review, it is worthy of note (a) that the agreement is not limited in point of time. It covers inventions which the employee has made or conceived or may at any time hereafter make or conceive . . .
(b) It is not limited to the subject matter to which the employee directed his attention when in the employ of appellee, but extends to any business, 'in which said company during the period of my employment by said company or by its predecessor or successor in business is or may be concerned.' In other words, if appellee's predecessor were engaged in any other business to which appellant's discovery might relate or its successor shall be or may be concerned, the contract applies.

"Upon the facts peculiar to this case we are convinced that those provisions of the contract which were limitless in extent of time and in subject matter of invention were contrary to public policy. Guth was a chemical engineer. He was more or less successful in research work, as is shown by the fact basis of this litigation. He was a research man prepared to devote his life to discoveries of value to industry. Under this contract he was, however, if he worked in another laboratory or for another manufacturer, required to assign his discoveries to appellee. This would effectively close the doors of employment to him. Until the end of the chapter he was compelled either to work for appellee or turn over the children of his inventive genius to it. Such a contract conflicts with the public policy of the land, which is one that encourages inventions and discourages the exclusion of an employee from engaging in the gainful occupation for which he is particularly fitted for all time, anywhere in the United States."

This is an extreme example of a clause in a patent assignment agreement which contravenes public policy and is therefore unenforceable. This principle serves as an effective restraint upon unreasonable employee invention assignment agreements. However, the courts have approved asignments of future inventions within the scope of the agreement if the time is definite and reasonable, such as for one year after termination of employment.

Secrecy Provisions

The enforcement of the secrecy provision requires entry into an area of great delicacy. Most employees, and particularly chemists, increase the general store of their knowledge in the course of their employment, and this broadening of their knowledge through experience they regard as part of their general knowledge and experience, to be used in any subsequent employment. Certainly, they are entitled to use such knowledge. However, they must be very careful in so doing not to use or divulge any of the trade secrets, know-how, or other secret information about their employer's activities or plants or methods of operation, since this is a definite violation of his property rights in these things.

An illustration of a chemist who went too far in this respect is the recent case of Monsanto Chemical Co. v. Miller (7). Monsanto over a period of more than 20 years had designed, built, and operated electrical furnaces for the production of elemental phosphorus. In May 1942, the employee, Miller, was engaged by Monsanto, and he continued in the employ of Monsanto, with an intervening break during World War II, until 1954, at which time he left.

In 1956 Miller was employed in a consulting capacity by Central Farmers Fertilizer Co. in research and study, looking towards the design and construction of an electric furnace for the production of elemental phosphorus. The plant which he assisted in designing for Central Farmers Fertilizer Co. was not then in the course of construction, but it was begun later.

It appeared that Miller had carried away from Monsanto about 102 physical drawings of the Monsanto plant, and he disclosed to Central Farmers, through background information acquired at Monsanto, how these drawings were to be applied in the design of the furnace for Central Farmers.

The court found that the complexity of the designing and the magnitude of it involved large amounts of money, engineering, experimentation, and information. Miller had executed an employment contract for Monsanto during his first period of employment but not during his second. It was not clear whether he was then obligated to keep confidential information concerning Monsanto's processes and other operations that he acquired during his second employment, but the court found that such obligation was at least implied, and that even though he were not subject to the terms of an employment contract, the nature of his employment was such as to subject him to the common law duty not to reveal engineering data designs, studies, or operating data confided to him while in the employ of Monsanto and which comprised trade secrets of Monsanto.

The court further found that Miller had deliberately acquired for future use and over a period of more than a year while in Monsanto's employ, sufficient information concerning the design, engineering, construction, operation, and capital and operating costs to enable himself and others to design and construct electrical furnaces similar to Monsanto's and that this information was in fact disclosed to Central Farmers. Accordingly, the court concluded that Monsanto was entitled to an injunction restraining Miller from using any of these trade secrets and made provision for an award of damages and costs to Monsanto.

Literature Cited[1]

(1) Bowers v. Woodman, 59 F.2d 797 (1932).
(2) Dinwiddie v. St. Louis & O'Fallon Coal Co., 64 F.2d 303 (1933).
(3) du Pont, E.I. de Nemours & Co. v. United States, 288 F.2d 904 (1961).

(4) Guth v. Minnesota Mining & Manufacturing Co., 72 F.2d 385 (1934).
(5) Heywood-Wakefield Co. v. Small, 87 F.2d 716 (1937).
(6) Houghton v. United States, 23 F.2d 386 (1928).
(7) Monsanto Chemical Co. v. Miller, 118 U.S.P.Q. 74 (1958).
(8) United States v. Dubilier Condenser Corp., 289 U.S. 178 at 188 (1932).

---

[1] Legal abbreviations are defined on page viii.

RECEIVED November 5, 1963.

5

Priority of Invention

ROBERT L. NIBLACK

Abbott Laboratories, North Chicago, Ill.

> In the United States priority of invention rather than early filing of an application for a patent entitles one to a patent. When several inventors claim coincident matter before the Patent Office, an administrative proceeding called an interference is used to determine priority. The inventor last to file his application must present corroborated proof of prior invention to overcome the priority presumption favoring the party who filed first. To build a strong priority position, one must first complete the invention as soon as possible and determine to the satisfaction of those skilled in the pertinent field, usually by testing, that it possesses the utility contemplated by the inventor. Second, the inventor should keep adequate records and develop corroborators who can testify to his activity.

In the United States priority of invention is important because the "first inventor" rather than the inventor first to file a patent application is entitled to award of a patent. An interference is a proceeding initiated by the Patent Office to determine which of two or more parties claiming the same or overlapping subject matter was "first inventor" and therefore entitled to an award of priority and issuance of a patent on the common matter.

Interference practice is derived from certain U.S. statutes (*18*) enacted pursuant to Article 1, Section 8 of the Constitution. Its underlying principle is that there can be only one valid patent for an invention.

An interference is one of the most complicated of legal procedures because it is governed by intricate, exacting rules, and a unique technical fact situation is ordinarily involved. This summary was written for

41

chemists and not for legal specialists. Some statements are made without qualification, although qualifications can be made to most any statement on interference practice. However, with inclusion of all the qualifications, chemists will, in the writer's opinion, learn very little.

Outside of academic considerations it is advantageous for inventors to have some understanding of the principles involved. Establishing legally correct routines for handling inventive situations and record-keeping can win interferences.

Basic Interference Principles

Invention for patent purposes consists of the elements of conception and reduction to practice. Conception is often defined as a complete mental realization of the invention. This is too narrow a definition, especially when considering chemical inventions (2). Preparing a new compound having no apparent use would ordinarily constitute conception of the compound. This laboratory preparation obviously goes beyond the mental stage of the invention.

Reduction to practice or completion of invention is accomplished "constructively" by filing a patent application (17). It is accomplished "actually" by physically completing the invention and determining its suitability for a designated purpose, usually by testing. Because constructive completion requires only filing a paper disclosure without time-consuming laboratory work or testing, advantages to be gained by early filing of a patent application are obvious.

The party who files his application in the Patent Office first, called the senior party (14), has a great advantage in an interference. The senior party is presumed to be first inventor. Those filing later (junior parties) must overcome this presumption by proof of earlier corroborated activity (19). The party proving first conception and first reduction to practice is always awarded priority. However, if the first conceiver is last to reduce to practice, he can prevail by proving "diligence" in reducing to practice during a period extending from sometime prior to the adverse party's entry into the field until completion of his invention (16). Diligence is a rather vague concept understood by few (13). Abandoning the interference subject matter in favor of a competing project is about the only activity that is definitely not diligence.

Corroboration is the major hurdle confronting an inventor attempting to prove actual reduction to practice. The statements of an inventor are not believable as an absolute matter of law. Everything to which he testifies must be corroborated by someone other than a coinventor having actual knowledge of the events (20). Ideal corroboration would be provided by someone unassociated with the inventor who completely duplicates the inventor's work or does all work under the direction of the inventor.

An interference is chronologically divided into four periods much like a court trial: a preliminary period; a motion period; a testimony period; and final determination. In the initial or preliminary period, notice of interference is given to all parties; affidavits may be required of junior parties to establish their *prima facie* right to be in the interference (*11*); and all parties are required to file binding statements (called preliminary statements) setting forth the earliest dates they later expect to prove. Ordinarily no dates earlier than those given in the preliminary statement can be later proved. The preliminary statement is designed to discourage fraud as a party might be tempted to stretch his dates or records upon realizing an adverse party was asserting earlier dates.

Patent Office Rule 222 provides for amendment of preliminary statements. However, this procedure is no substitute for careful preparation. One should not give up the search for earlier dates and records but should consider the Patent Office Board of Patent Interferences' recent decision (*12*) denying a motion to amend because of no showing that a proper degree of effort and care was exercised in preparing the original statement.

The motion period prepares the proceeding for testimony. During this second period, parties may move to amend counts, to bring in additional subject matter, or to limit the subject matter in issue. Motions may be made to dismiss by showing reason why there is not or should not be an interference, or the order of the parties may be reversed because of earlier-filed applications disclosing the invention in conflict. Upon setting of the motion period, each party gains the right to inspect previously secret applications of other parties.

The testimony period follows the motion period. After testimony periods have been set, each party may inspect the preliminary statements of others. Testimony is presented by depositions which are statements sworn to under oath with right of cross-examination by adverse parties. The most junior party takes testimony first followed by more senior parties in order of their filing dates. If a junior party fails to take testimony, the senior party may ask for judgment on the record to end the interference on the basis the junior party did not overcome the presumption of first inventorship which favors the senior party (*19*).

In the fourth and final portion of the interference, the record of each party is filed with the Patent Office; oral arguments may be presented before a three-man Patent Office board; a decision is rendered by the board, and appeal or re-opening of ex parte prosecution takes place.

Reduction to Practice

It is obvious that an inventor fighting for an award of priority in an interference will strive to establish a reduction by going back to his earliest activity. Ideally, reduction to practice is established by proving commercially successful use or testing under actual service conditions. Because such activity usually takes place years after bench testing and filing of patent applications, the inventor looks to more questionable crude, initial activity to prove his case. Therefore sufficiency of laboratory testing is a critical issue in most interferences.

The three prerequisites of patentable invention are novelty, unobviousness, and utility. Proof of utility is basic to reduction. As a general rule, a substance is reduced to practice when it is actually produced unless its usefulness is not apparent from its ingredients, chemical structure, or manner of production, in which case the material's utility must be demonstrated by proper tests (*15*).

An inventor must be reasonably certain that any testing results can be correlated back to the utility he had in mind at the time of testing. To find the utility contemplated by the inventor, the Patent Office and the courts look to any utility given by the count (the patent claim common to the interfering parties): (*a*) If a use is given in the claim, only proofs relating to that use will be examined; (*b*) if the count doesn't specify a use, the patent specification is examined to find what use was contemplated; (*c*) what is found in the specification is tempered by behavior and records of the inventor and those around him.

As an illustration of this procedure, a recent case (*7*) involved the count "stabilized polyethylene compositions comprising a normally solid polymer of ethylene and a stablizing amount" of a specified material. The laboratory work testified to was limited to milling batches of polyethylene at high temperatures, pressing them into plaques, and exposing the plaques to several tests. Because making stable plaques was obviously not the utility the inventor had in mind at the time of testing, the court was presented with the problem of deciding whether these tests could be correlated with whatever utility the inventor then envisioned. A clear utility not being expressed in the claim, the court looked to the specification which stated the object of the invention was to produce compositions having odor, color, and electrical stability. Testimony of the inventor's associate established that it was his responsibility to develop the product as a wire coating.

The court looked at the count and found no clear limitation regarding utility. After examining the specification which mentioned cosmetics, food packaging, and wire insulation, the court was still up in the air. Then it looked at the project aim stated by the corroborating witness and set the utility standard by the testimony. Insulating of electrical wires was the utility actually contemplated by the inventor at the time of testing, and it was required that his asserted reduction to practice prove the suitability of the defined composition for that purpose. Because the results obtained from testing of plaques could not be correlated to coating electrical wires, no reduction was proved.

In determining the utility contemplated at the time the invention was made, an inventor cannot be forced to prove a utility not given in his application. The old "ultimate use" doctrine which forced extension of utility past that stated in the specification has been buried by recent decisions. An inventor's ostensible objective governs. For example, tests in animals are adequate if a pharmacological utility is asserted, and the inventor cannot be forced to extend testing to humans on the basis that such is the ultimate use of any therapeutic (*1*).

Proving Utility

Assuming the pertinent utility is clear, what kind or degree of testing, if any, is required to prove that utility?

To prove utility and consequently establish an actual reduction, facts must be presented that will convince the man skilled in the field that the invention will function satisfactorily for its intended purpose (*8*). We have just explained how the intended purpose is determined. To determine what those skilled in the particular specialty involved would require to be convinced the invention would function as intended, three things are considered: (*a*) published literature on the subject or analogous subject; (*b*) what did the inventor think was necessary to establish utility; (*c*) any special nature of the intended use. If the literature indicates the invention is obviously useful for the intended purpose, testing may not be required (*10*).

Although, in theory, testing is required only in cases where those skilled in the art would require it to be reasonably sure of success for the intended use, certain subject matter almost always must be tested. A leading example of this is airplane inventions. Actual flight tests are generally required (*6*). Complex mechanical structures are another example. Common sense brings others to mind. For example, an invention intended as a foolproof safety device would probably always require testing under actual use conditions.

The prudent inventor will generally consider testing necessary. It has been held necessary to test devices as simple as a mattress handle (*5*), and hand-holding of a hearing aid rather than using a head band during testing has been held inadequate (*9*).

Outside of a few areas, such as aircraft and safety devices, the inventor himself may determine the degree and kind of testing needed. If he conducts a test and at time of testing concludes the invention will satisfy the utility he envisions, that is ordinarily sufficient. If he admits to inadequacy of his testing, that may be fatal to a later contention of success (*4*).

In other words, in many cases the inventor sets the standard as to what those skilled in the art would require. Obviously the inventor can't be an erratic genius, nor can he circumvent precedents or common sense.

If his standard is reasonable the inventor's own attitude controls adequacy of testing (*13*).

Returning to the polyethylene stabilization case (*7*), the inventor stated in his testimony that plaque testing done years before was sufficient, but at the time he reported these results, he wrote the "encouraging results . . . warrant further investigation." Because of this statement he was precluded from later saying the laboratory tests were as good as tests under actual service conditions. His own records indicated plaque testing was a preliminary thing.

Tests, to be adequate, must establish that the invention worked satisfactorily under service conditions; that the laboratory tests duplicated actual use conditions; or that test results can be directly correlated to results obtained under actual use. If a test is recognized by those in the field as standard, and there is correlation between laboratory results from this test and those resulting from actual usage, this test can be used to establish reduction (*8*).

Whether a particular in vivo animal test is sufficient to establish usefulness of an invention in humans depends on determination of what is a standard test animal. In the Hartop case (*8*), the inventors, to demonstrate safety of a thiobarbiturate anesthetic composition, submitted rabbit test results coupled with evidence that rabbits were standard tests animals for this area. The court held the specification indicated that although human therapy was contemplated the rabbit tests were sufficient to prove usefulness of the composition for the inventors' purpose. Tests under actual conditions of contemplated use—that is, in humans—were not required because one skilled in the art would accept the tests as indicating it to be reasonably certain that the invention would have the utility alleged.

In summation, the following points should be observed or at least considered by inventors. Although this is not an exhaustive listing of interference procedure, each point is substantially under the control of an inventor, and proper attention to them will strengthen his priority position.

  1. File patent application as early as possible. The senior party has the battle half won.

  2. Don't make important research projects a hobby of long standing. Complete the work and go on to something else. Periodic renewal and abandonment of a project is the surest way to destroy diligence.

  3. Keep complete reasonably up-to-date records.

  4. Develop corroborators. Conception proofs require the testimony of other persons to whom the concept has been explained. Reduction-to-practice proofs require corroborators who possess personal knowledge of any activity to which they testify. Short-cut procedures, such as

mailing sealed documents to oneself, do not provide corroboration. Even the act of witnessing notebooks and affixing the "read and understood" stamp mark does not provide corroboration as to a reduction to practice (although such records may aid in refreshing the memory of a corroborator who has actually seen the work performed).

5. Don't be hypercritical of research results. An inventor's opinion of the success of a project as evidenced by his writings made at the time is very important. He can't state failure originally and then attempt to claim success at a later time when confronted by overlapping work of an adversary.

6. Carefully consider who should be named inventors on a patent application. The intellectual contribution of a worker determines his right to be named inventor, not the number of hours spent or the amount of administrative supervision he gives. An inventor cannot be a corroborator. Therefore, if everyone associated with the invention is named inventor, there are no corroborators.

7. Generally assume successful testing is required for reduction to practice. Choose tests that are considered standard; and choose those that have as direct correlation to the inventor's stated object and utility as possible.

8. Check patent applications to ensure use of utility statements that can be supported by early tests. Don't expect a 30-second, one-time test to support an application that states the purpose of the invention is to provide a longer-lived, more stable and durable material (3).

9. When filing a preliminary statement or taking depositions becomes necessary, carefully review records at the outset. An inventor will probably be required to live with dates originally asserted. As a horrible example, testimony proving early activity is excellent to prove a later date asserted in a preliminary statement, but it will not entitle an inventor to the earlier date.

Literature Cited[1]

(1) Archer v. Papa, 121 U.S.P.Q. 413.
(2) Bac v. Loomis, 117 U.S.P.Q. 29.
(3) Conner v. Joris, 113 U.S.P.Q. 56.
(4) Duddy v. Solomon, 113 U.S.P.Q. 294.
(5) Fridolph v. Bechik, 69 U.S.P.Q. 128.
(6) Gaiser v. Linder, 117 U.S.P.Q. 209.
(7) Harding v. Steingiser & Salyer, 138 U.S.P.Q. 32.
(8) Hartop & Brandes, In re, 135 U.S.P.Q. 419.
(9) Koch v. Lieber, 61 U.S.P.Q. 127.
(10) Kyrides v. Bruson, 41 U.S.P.Q. 107.
(11) Miller, Ex parte, 124 U.S.P.Q. 419.
(12) Moler & Adams v. Purdy, 131 U.S.P.Q. 276.

(13) Morway, Beerbower & Zimmer v. Bondi, 97 U.S.P.Q. 318.
(14) M.P.E.P. 1102.1.
(15) Rodin v. Spalding, 132 U.S.P.Q. 285.
(16) 35 U.S.C. 102.
(17) 35 U.S.C. 102(e).
(18) 35 U.S.C. 135(a).
(19) U.S. Patent Office, Rule 252.
(20) Wever v. Good & Putzrath, 129 U.S.P.Q. 32.

---

[1] Legal abbreviations are defined on page viii.

RECEIVED October 31, 1963.

6

Formal Documents of the U. S. Patent System

PAUL D. BURGAUER

Abbott Laboratories, North Chicago, Ill.

> The various documents an inventor must sign when he files an application for a patent are explained. Included are the documents required by the Patent Office before an application is accepted and those which may be necessary during the subsequent prosecution of the application in the U. S. Patent Office. Among others, the oath, power of attorney, petition, preliminary statement, and affidavits are discussed specifically.

Many chemists for whom I prepare patent applications have asked why they must sign any of the formal papers, whether they have to read what they sign, and what their signatures will accomplish. These interrogations are based on anything from interest or curiosity to implied annoyance with such formalities.

The first and most basic paper pertaining to patents which needs the inventor's signature is the oath. This is a required instrument without which no patent application can be filed, since the Patent Office rules prescribe that a patent application to be examined by the Patent Office must be accompanied by an oath, a petition for the grant of the patent, and the filing fee. The oath is, in most cases, a formal document which states the country of which the applicant is a citizen [it has been held that "the inventor intends to become a U.S. citizen" is unacceptable (2), but the statement that "he is a citizen of no country" was accepted (13)]. It also states that the applicant believes himself to be the first and original inventor of the improvement described and claimed in the attached specification; that he does not know and does not believe that the same was ever known or used before his invention thereof or patented or described in any printed publication in any country before his invention thereof or more than one year prior to the application, or in public use or on sale in the United States more than one year prior to the application; that the invention has not been patented in any country foreign to the United States on an application filed by him or his legal representatives in the 12 months prior to this application; and that no application for patent on the invention has been filed by him or his representatives in any country foreign to the United States except as follows.

The phrase, "that no application has been filed in any country foreign to the United States except as follows" is important for patent applications which are filed as counterparts to a first-filed foreign application and is directed mainly to inventions made abroad which have been first filed in a country foreign to the United States. It is also of utmost importance when it appears in a continuing application—that is, an application that had to be refiled for one reason or another but which had also been filed abroad during the time elapsed between the first U.S. application and the date of the new oath. The paper is then called "an oath for a continuing application."

Patent applications must be filed within a reasonably short time after execution of the oath. The court accepted an oath that was five weeks old (8) but refused an application, which was filed promptly after execution of the oath, because filing became effective only when the filing fee was submitted five months later (11).

The oath is an integral part of the application (specification and claims) to which it refers, and it must be attached thereto. After executing the oath, no changes can be made. The courts are so strict on this point that an applicant lost his right to his filing date because he signed the oath attached to the application and in the cover letter with which he returned it to the attorney he insisted that the attorney make some minor changes before filing (1).

Oath

Since the oath is the form that is most frequently submitted to researchers for signature and since it is a relatively lengthy formal document, it appears advisable to analyze its content in more detail.

The wording in the oath is actually based on the patent statute which specifies that a person is entitled to a patent for an invention, plant, or design unless (35 U.S.C. 102):

1. The invention was known or used by others in this country, or patented or described in a printed publication in this or a foreign country before the invention thereof was made by the applicant; or

2. The invention was patented or described in a printed publication in this or a foreign country or in public use or on sale in this country more than one year prior to the date of the application; or

3. The applicant has abandoned the invention; or

4. The invention was first patented or caused to be patented by the applicant or his legal representatives in a foreign country prior to the date of the application in this country on an application filed more than twelve months before the filing of the application in this country; or

5. The applicant did not himself invent the subject matter; or

6. Before the applicant's invention, the invention was made in this country by another who had not abandoned, suppressed or concealed it; or

7. The invention was described in a patent granted on an application for a patent by another filed in the U.S. before the invention by the applicant.

The last condition, of course, is one that cannot be sworn to, and therefore the corresponding wording is absent from the oath: the applicant cannot know whether his invention has been filed as a patent application by another so long as that other application has not resulted in a printed publication.

For the sake of completeness, it should be mentioned here that different forms are prescribed for an oath not accompanied by the application; an oath for more than one inventor; a supplemental oath covering subject matter not originally claimed in the application attached to the original oath; an oath for a continuing application containing or not containing additional subject matter; or an oath executed by a guardian or administrator in event of the inventor's incapacity or death. In all instances, the legal effect is the same as in the original oath and, therefore, the content does not vary substantially therefrom.

Power of Attorney

Although the law provides that the inventor himself may file an application on his invention, in practice most inventors seek the assistance of a patent agent or attorney. This then requires the inventor to sign a power of attorney, giving the person or persons named therein the right to represent him at the Patent Office. Either the inventor or the assignee can empower an agent or an attorney. The application may even be filed by the assignee if the inventor refuses to sign the oath to the application. But in this event, the assignee must prove by affidavits and the like, to the satisfaction of the Commissioner of Patents, that he is the owner of the invention and that the inventor refuses to sign, before the application can be filed (6). And even then, the application must be filed in the name of the inventor.

Where the power of attorney is given to a law firm, changes of personnel within that firm will not change or nullify this power. Where the power of attorney is made out to individual members of a patent staff, complications may arise when that staff changes. For instance, if one member of the staff is replaced by someone new, the new member has no power, while the old member is still empowered. Only the applicant himself can revoke that power. The most common practice in research organizations that have their own house counsel is to give the power of attorney to the patent counsel or the head of the patent staff, who, in turn, can delegate his power to others. The power given to the principal attorney—that is, the attorney to whom this power is made out originally—becomes void upon the death of that attorney, and with this nullification all powers delegated by him are nullified. Thus, it is recommended that at least two principal attorneys be named in the power of attorney, indicating with which one correspondence is to be conducted. Then, in case of the death of one, the other can still prosecute the application and/or delegate this power to others. But the Patent Office will not undertake correspondence with more than one attorney at a time. Actually, no immediate danger occurs even if the power is given to only one attorney and he withdraws from this function for any reason, since the Patent Office will put on file any paper submitted by another registered attorney or agent and will enter that paper upon ratification of a new power of attorney to the person who submitted it. In other words, if the attorney handling an application dies and one of his delegates files an amendment in timely fashion, this amendment will be entered when, subsequently, a new paper is submitted to the Patent Office ratifying a power of attorney to that agent or attorney. Obviously, the power of attorney must be given to an agent or attorney registered in the Patent Office. To become registered as attorney or agent requires the passing of an examination and the issuance of a certificate of registration by the Patent Office.

As mentioned before, the inventor himself has the right to prosecute an application. If the Patent Office receives a response signed by the inventor (in an application in which there is an attorney of record), any amendment therein will be entered, and the response will be acted upon. However, in this action the Patent Office will call attention to the rule which specifies that correspondence will be held with only one person.

Where there is more than one attorney or agent named in the power of attorney, the one latest appointed will be addressed, in the absence of other requests. It is recommended that the power of attorney include a Washington lawyer as an associate or that the principal attorney delegate power to a Washington colleague since, in emergencies, the Patent Office will call such a Washington associate if he is on its records as such. In the alternative, the examiner should be given permission to get in touch with the appointed attorney by a long-distance, collect telephone call.

Petition

Another document that must be filed with the patent application is the petition. This is also a formal document stating that the applicant is a citizen of a certain country and that he prays that a patent may be granted to him for an improvement in such-and-such as set forth in the specification to which said petition is attached.

Both the petition and the oath must be signed, and the oath must be sworn to and notarized. This is one reason why the so-called "short form" or "single-signature form" has been approved by the Patent Office. In this short form, the wording of the petition and the oath are combined and therefore only one signature and one notarization is required. The same form also names the attorney and confers the power on him. The content of this short form is the same as that of the separate documents: the applicant states that he is a citizen of country X; that he believes he is the first and original inventor; that he appoints an attorney or agent to prosecute the application on his behalf; and that he prays a patent may be granted to him. This form is "short" only in the sense that the petition, the oath, and the power of attorney are combined into one document.

Preliminary Statement

Under our Patent Office law, a patent may be issued only to the first and original inventor or inventors. Where two or more inventors disclose and claim the same patentable invention either in a plurality of applications or in an application or a patent, the Patent Office will give the parties an opportunity to contest priority of invention by way of an interference. An interference is a contest set up by the Patent Office to determine the rightful inventor of common subject matter contained and claimed in two or more applications or patents. In the process of determining the first inventor, another legal document is required—the preliminary statement. This document states, above the inventor's notarized signature, the various dates on which the various steps were carried out which led to the reduction to practice of the invention. The dates required are those on which the first drawing of the invention, the first written description of the invention, and the first disclosure to another person were made. Further, it requires the date when the first actual reduction to practice occurred and the date after conception from which reasonable diligence toward reduction to practice began. Once this document is approved by the Patent Office, the dates therein can be changed only by motion and with a satisfactory showing that correction is essential to the ends of justice. All alleged dates in the preliminary statement must be provable by original, signed, dated, and witnessed documents to meet the tests applied during trial. It is thus of utmost importance that these dates be carefully established and be supportable.

This brings me to another point that may not have been sufficiently stressed before—keeping good records. Your scribbled notes on a loose leaf are worthless when it comes to proving any one of the above dates, but your well-kept laboratory notebook with regular entries which are signed, dated, and witnessed by proper persons will be of immense help to your agent or attorney when the time comes to cite or to substantiate the dates alleged in the preliminary statement. Although your laboratory journal has no evidentiary character unless corroborated, the time may come when the journal entries may be the document with which an interference between you and an inventor not connected with your company is won or lost (5).

In connection with the notebook, which may be of significance, another valuable statement can be the conception records or patent proposals or disclosure sheets, or whatever other name is given to the formalized documents submitted by laboratory workers to a patent attorney or agent. The signature thereon may also become of great legal significance. Here, as well as in notebooks, a clear and concise description of the invention is important, but in contrast to the notebook, it can be more clearly directed to the point of invention than can the lab notebook which only serves to record step-by-step procedures, failures, and observations of no future importance. The conception record should:

Show the disadvantages of the prior art, process, or apparatus.
Show how the invention overcomes such disadvantages.
Describe the details of the inventive procedure.
Describe additional advantages attained with it.
List precisely what is new.

A record like this is of immense help to patent attorneys or agents in preparing patent applications, but also, if properly dated and witnessed, it may establish a provable date in an interference procedure. Thus, again, a routinely applied signature and witnessing procedure may convert a conception record, as well as a lab notebook, into a document of high legal value.

Affidavits

Another document that requires the signature of the researcher is the affidavit. Affidavits are sworn papers submitted to the Patent Office to prove or disprove facts. Examiners rely on these papers to deny patentability of the invention. If an affidavit, submitted during the prosecution, convinces the examiner of the applicant's position, and the affidavit later proves to be false or falsified, the patent is invalid (9, 10). Thus, utmost care should be taken in preparing an affidavit.

Essentially, there are two kinds of affidavits—those designed to establish conception and reduction to practice prior to the effective date of a cited reference (Rule 131) and the affidavit under Rule 132 to show the value of the invention in the light of the prior art. These affidavits are referred to by the rule numbers in accordance with the "Rules of Practice of the United States Patent Office."

The Rule 131 affidavit requires the affiant to make an oath to facts showing completion of the invention before the effective date of a cited reference. The effective date in the case of a foreign patent or a domestic or foreign publication is the issue date thereof. But when a U.S. patent is cited, its filing date must be antedated. This, of course, is true only when that reference describes the invention. The showing of facts must establish reduction to practice prior to said date or conception of the invention coupled with due diligence from the conception date to a subsequent actual or constructive reduction-to-practice date. The term "constructive reduction to practice" refers to the filing date of the application in the Patent Office. For the period following the date of actual reduction to practice, proof of diligence is not needed but the date of actual reduction to practice must be provable. Original exhibits of drawings or other records must be attached to this type of affidavit, or their absence should be explained to the satisfaction of the Patent Office. Again, the notebook properly kept, dated, signed, and witnessed is of great help. If necessary, this Rule 131 affidavit can be made by someone other than the inventor when it is satisfactorily shown why it is not executed by the inventor or applicant. Of course, where an invention is anticipated by an issued U.S. Patent which claims that subject matter, this affidavit will not overcome the reference. An interference is necessary, but this is only possible if the claims are copied within one year of the date of the issued patent.

A Rule 132 affidavit is most commonly used to show operability, an improved result in yield or time, more economical conditions, or the improved properties of the result. No weight is given by the Patent Office to an affidavit which shows by expert opinion how the claim or the disclosure should be construed, since the Patent Office takes the position that the examiner knows best how to interpret the language used. Also, little or no weight is given to an affidavit showing commercial success of an inventive product if that is the only argument available for patentability. However, along with other evidence, it may turn the tables in a tight situation. This Rule 132 affidavit usually contains two major portions: the first portion asserts the qualifications of the affiant as an expert in the field, and the second portion shows a comparison between the inventive process or product over the process or product of the prior art. The affiant is not necessarily the inventor or applicant; in fact, quite frequently it is desirable to ask another experienced person in a given field of research to make such an affidavit, since this third person may be considered a person disinterested in the outcome of the particular prosecution difficulty. His opinion therefore may carry more weight in showing the real merit of the invention (*3, 4, 7, 12, 14, 15*).

Assignment

The only other paper routinely signed by chemist-inventors is the assignment. However, signing your name to the assignment form provided by your employer is a mere formality, since it does nothing more than entitle your employer to all rights of your invention, to which he is already entitled if you have been hired to invent.

Literature Cited[1]

(1) Ames v. Lindstrom, 1911 C.D. 69.
(2) Benecke, Ex parte, 1907 C.D. 66.
(3) Brown, Ex parte, 2 U.S.P.Q. 342.
(4) Davidson, In re, 12 F.2d 814.
(5) Field, J.P.O.S. 43, 591 (1961).
(6) Gray, In re, 115 U.S.P.Q. 80.
(7) Gustavson, Ex parte, 14 U.S.P.Q. 332.
(8) Heinze, Ex parte, 1919 C.D. 67.
(9) Kleinman v. Betty Dain Creations, 89 U.S.P.Q. 404.
(10) Levine v. United States, 77 U.S.P.Q. 124.
(11) Miller, Ex parte, 1908 C.D. 286.
(12) Reifsnyder, Ex parte, 1 U.S.P.Q. 41.
(13) Rhodes, Ex parte, 1903 C.D. 257.
(14) Schluchter, Ex parte, 49 U.S.P.Q. 332.
(15) Teppema, Ex parte, 18 U.S.P.Q. 289.

[1] Legal abbreviations are defined on page viii.

RECEIVED October 31, 1963.

Meaning and Interpretation of Chemical Patents

SYDNEY G. BERRY

Berry and Crews, 25 W. 43rd St., New York, N. Y.

> The subject of chemical patents is introduced by considering the structure of the patent specification, the terminology peculiar to patents in chemical and nonchemical fields, and the effect of these factors upon validity. Also, requirements for disclosure in the specification of chemical patents are discussed, including the showing needed to establish utility, especially with regard to patents on drugs. To illustrate the importance of some of these matters when it comes to a suit on the patent, such a suit is followed through the trial and appeal courts up to and including the U. S. Supreme Court, with a discussion of some other aspects of the patent law that are brought out in the suit.

In 1960, approximately 20% of the patents filed and issued were classified as chemical; this included patents in the field of metallurgy. On the other hand, when it came to appeals and interferences, no less than 50% of the cases involved chemical patents. One can only speculate on the reasons for this. Very likely the relatively greater number of interferences has a causal relation to the great amount of research that is carried on by chemical firms, some of whom work along quite parallel lines. As to the increased number of appeals, there is no doubt that this is a product of the more complicated prosecution of chemical cases than obtains with mechanical or even electrical cases. In effect, the evidence for this is largely the subject of this paper.

Recently the writer was presented with a facsimile copy of U. S. Patent No. 1, granted July 31, 1790, which bears the signatures of George Washington and Edmund Randolph, then president and attorney general, respectively, of the United States. It turned out to be a chemical patent, claiming an apparatus and process of making pearl ash, which is an old term for potassium carbonate.

At the outset, a word may be said concerning terminology found in patent specifications. Many look upon patents as having a terminology of their own. No doubt there is a grain of truth in this, because (*1*) there is need for greater exactitude and greater inclusiveness in describing technical matters; (*2*) frequently the art is so new that no proper terminology is available; and (*3*) a patent specification is both a technical and a legal document.

This brings us to a consideration of the official rules issued by the Patent Office (*17*) insofar as they relate to the content of the patent specification. We may quote Section (a) of Rule 71:

"The specification must include a written description of the invention or discovery and of the manner and process of making and using the same, and is required to be in such full, clear, concise, and exact terms as to enable any person skilled in the art or science to which the invention or discovery appertains, or with which it is most nearly connected, to make and use the same."

This specification is not addressed to the general public but rather to those skilled in the art to which the invention or discovery appertains. Seemingly, it is *carte blanche* for the inventor or his patent attorney to use the language of the technology without restraint, and undoubtedly many a patent specification is so written. Yet it should be borne in mind that when the patent is sued upon, the judge who must interpret it is nearly always a layman, as far as the art is concerned to which the patent relates. It is fairly obvious that if patent specifications are couched in such abstruse technical terms that the court cannot fully understand them, even with the aid of experts, your chances of succeeding in enforcing your rights and having the patent declared valid are seriously diminished. Therefore, while addressing the patent specification to the people skilled in the particular art to which it appertains, it behooves one to do so in a manner as understandable as possible. For this reason alone, a patent specification may well be more readable than the ordinary scientific paper covering the same subject matter, notwithstanding the fact that it has a phraseology all its own.

Section (b) of Rule 71 provides that:

"The specification must set forth the precise invention for which a patent is solicited, in such a manner as to distinguish it from other inventions and from what is old. It must describe completely, a specific *embodiment* of the *process, machine, manufacture, composition of matter* or *improvement* invented, and must explain the mode of operation or principle whenever applicable. The best mode contemplated by the inventor of carrying out his invention must be set forth."

Also, Section 100 of the Patent Code provides:

"(b) The term 'process' means process, art or method and includes a new use of a known process, machine, manufacture, composition of matter or material."

These sections have particular significance to the drafter of specifications of chemical and metallurgical patents because, following the statute, they enumerate those categories of inventions in which chemical and metallurgical patents are contained—namely, processes, compositions of matter, and uses.

In Section (b) of Rule 71, the words "specific embodiment" are important, since if a chemical invention is sought to be patented, it is necessary to give one or more specific examples of it. A description in general terms only, formerly permitted, will no longer suffice.

Section (c) of Rule 71 requires that when an improvement is sought to be patented—and this may be assumed to take in the vast majority of patents—the description is to be confined to the specific improvement and "such parts as necessarily cooperate with it or as may be necessary to a complete understanding or description of it."

We have now seen what the categories are for patentable inventions and which contain those patents which relate to chemistry and metallurgy. Let us suppose then, that as an inventor, you carefully follow the provisions of Rule 71, and the Patent Office approves your specification as to form, and after a thorough search of the art, finds your invention to be novel, do you then get the patent? The answer is "not necessarily," because you still must surmount the hurdle provided by Section 103 of the patent statutes, which reads in part as follows:

"A patent may not be obtained though the invention is not identically disclosed or described as set forth in Section 102 of this title, if the differences between the subject matter sought to be patented and the prior art are such that the subject matter as a whole would have been obvious at the time the invention was made to a person having ordinary skill in the art to which said subject matter pertains."

Thus, to be patentable, your new chemical or metallurgical process, composition or alloy must not only have this element of invention—most difficult to define—but as a patent specification draftsman, you must take pains to present your invention in the manner which will best bring out this element of invention as distinguished from what is the result of "mere skill of the calling." Thus you have, most of the time, a scientific paper which is a legal document and which is also, in a rather subtle and unobtrusive sense, a sales argument. One example of this is the frequently found expression "most surprisingly I found so-and-so to be the case." Or "it was not to be expected that compound A would react with compound B to give compound C." And then, you must be careful not to say anything in such a way as to give rise to a presumption of abandonment of any part of your invention.

This point may be illustrated by a case decided some years ago (1941) in the United States District Court for the Northern District of Illinois. The case (*19*) is a bit extreme but admirably illustrates the point. It had to do with a certain use of mineral talc, and for some reason, the patenteé specified "French talc" without reservation. The defendant had substantially the same process, only he used "California talc." The court held that the patentee had limited himself to French talc, and even though French talc and California talc were pretty much the same (the plaintiff said "talc is talc"), there was no infringement. The case thus presents a lesson to the specification draftsman, and that explains a word that you will find so often in patent specifications—namely the word "preferably." In this instance, the patentee should have specified that he preferred the use of French talc, although his invention could be practiced to a satisfactory degree if some other talc had been used.

Further, you must be careful to put into your specification a sufficient foundation for what you wish to claim as your invention.

The following example (*4*) involving an invention of the late Carlton Ellis, is in point. Mr. Ellis was a prolific inventor in the chemical field, who, during his lifetime, took out over 700 patents. His first claim to fame resided in his invention of a paint and varnish remover consisting of a paint solvent and a dissolved wax, which, when the mixture was applied to the paint, prevented the solvent from evaporating until it had accomplished its purpose in loosening the paint. In a subsequent application for patent, he claimed the use of a finish-remover in which the active ingredient was a "ketonic derivative of a cyclic CH_2 hydrocarbon." The Patent Office refused to allow a claim of this scope, but on appeal to the Court of Appeals of the District of Columbia, the court reversed the Patent Office on the ground that Ellis had disclosed as many as 20 substances coming under this designation (*4*). On the other hand, when two other inventors—namely, Dosselman and Neymann (*3*)—came along with an application for a somewhat different type of finish-remover which called for a ketonic composition for removing surface finishes, their appeal for the grant of a broad claim was turned down by the same court on the ground that their disclosure of a single ketone— namely acetone—did not entitle them to claim a ketonic finish softening material broadly.

The decision was based upon the old Supreme Court "Incandescent Lamp Case," decided in 1895 (*16*). The Dosselman and Neymann case marked the beginning of a long series of cases involving the adequacy of disclosures in chemical applications, which have caused the chemical inventor and his attorney to multiply examples in order that there may be avoided the familiar rejection: "The claims are rejected as broader than the invention." Space does not permit a discussion of these many cases, which at times put an exceedingly heavy burden upon the inventor, but one may conclude with mention of a recent one. In this case (7), the

Court of Customs and Patent Appeals has perhaps liberalized the doctrine of the previous cases considerably, by holding: "It is manifestly impracticable for an applicant who discloses a generic invention, to give an example of every species falling within it, or even to name every such species." Further the court states: "It is sufficient if the disclosure teaches those skilled in the art what the invention is, and how to practise it."

In general, a broad invention will require many more examples than a narrow invention, and if, in your perusal of a chemical patent, for instance, you are surprised at the wealth of specific examples, it is because the patentee is endeavoring to form as broad a base as he can for his broad claims.

A still further requirement that in recent years has assumed greatly increased importance is that of establishing utility of a particular chemical compound or alloy.

Thus, in the year 1950, an inventor named Tolkmith (*15*) presented a patent application claiming a certain methane phosphonic chloride. The only utility stated for the compound was that it was of value as an intermediate for the preparation of more complex phosphorus derivatives and as a constituent of a parasiticide. The Patent Office Board of Appeals in its decision, held that the applicant's showing of utility did not comply with Section 112 of the patent statutes requiring that the manner and process of making and using the invention be described in such exact terms as to enable one skilled in the art to make and use the same. The board went on to say:

"In our opinion, the minimum requirement to satisfy Section 112 and Rule 71 on this aspect of the invention would be a specific embodiment of the composition in a parasitical composition with a disclosure of how it is to be applied, and to what parasite."

Since this decision was handed down in 1954, the Patent Office has placed considerably greater emphasis on a showing of utility, a requirement specified by the statutes in Sections 101 and 112. In 1960 the Court of Customs and Patent Appeals handed down a decision in an important case (*14*), in which the subject of utility was exhaustively treated. Seemingly it has considerably liberalized Patent Office practice on the subject. The Nelson case held essentially that it was sufficient for the applicant to have stated that the new chemical compound, a steroid, was useful as an intermediate for the preparation of other steroids which themselves were alleged to have valuable therapeutic properties, all without submitting proof that such therapeutic properties existed. Seemingly the section of the statute most involved was 112, which requires the applicant to give a written description of the manner and process of making and using the thing claimed. The subject of utility is also much involved when the compound sought to be patented is a drug or medicine or a method of treating disease in humans. Here we find the District Court of the

District of Columbia, in 1957, in the case of Isenstead v. Watson (*8*), setting forth that the Patent Office should be very careful, and perhaps even reluctant, to grant a patent on a new medical formula until it has been thoroughly tested and successfully tried by more than one physician. However, we find a rather different point of view expressed by the Court of Customs and Patent Appeals (*10*), which held:

"There is nothing in the patent statutes, or any statutes called to our attention, which gives the Patent Office the right or duty to require an applicant to prove that compounds or other materials which he is claiming and which he has stated are useful for 'pharmaceutical applications' are safe, effective, and reliable for use with humans."

Although only the results of animal experimentation were submitted as evidence of the effectiveness and reliability of the claimed compound, the court held that this was sufficient.

This brief discussion of utility should not be closed without mention that it is presently in a state of flux and that decisions in future cases may be expected which conceivably may modify substantially the doctrine as it is presently constituted.

We may now return to the Rules of Practice, and specifically Rule 75 (*17*), which states in part:

"The specification must conclude with a claim particularly pointing out and distinctly claiming the subject matter which the applicant regards as his invention or discovery."

The claim or claims must be read when you are required to know what the scope of the invention is. Even if you are not under the necessity of reading the claims, you should understand their function, since practically every part of the specification has been drafted with the claims in mind and to provide a foundation therefor. As an example of a claim, the following one, descriptive of a rather famous early invention (*20*) may be cited:

"A pyrophoric alloy containing cerium, alloyed with iron, substantially as and for the purposes described."

Now, because the right words are so important, especially in the claims, it is an established rule that the specification is to be considered the dictionary for the claims, and this assumes increased significance when the art is so new that the necessary terminology is scant.

In any discussion of claims, the terms "genus" and "species" will most likely arise. Perhaps a simple example will suffice to illustrate these terms. Suppose I invent a novel process of bleaching pulp employing a halide salt; a claim to a method of bleaching by the action of the halide salt would be properly termed a "generic claim" or "genus." On the other hand, if I then claim a chloride salt, this would be a species coming under the genus of halide salts. So would a bromide or iodide salt. In this example, let us suppose that fluoride salt would be inoperative. This would preclude me from claiming the genus, or if I did, it would be grounds for holding such a claim invalid. At the present time you may claim as many as five species, provided you also claim a genus which covers all the five species. We will have an opportunity to consider this subject again when we discuss a specific patent and the litigation involving it.

Having attempted to give some notion of the problems of the specification draftsman and particularly his concern with an adequate disclosure to the public—his specific examples, let us say—and of his equal concern that nothing be dedicated to the public that is properly part of the invention, let us consider some of the words that the patentee uses in the pursuit of these ends.

We have already considered "preferably," which, while pointing to the element of desirability, yet also has in it the element of reservation. It is, therefore, a word much used in the patent specification.

At the opposite pole is the word "essential." This word must be used with care, since the commitment here is final and cannot be undone by any amount of subsequent explanation.

The word that perhaps has the greatest currency, is "substantially." It is used because, in the nature of things, there are few absolutes. It is used in many ways. Thus, one may call for a temperature "substantially" above the boiling point of water at atmospheric pressure, or he may call for this or that ingredient in "substantial" amounts or proportions. The courts have had many occasions to comment on the use of the term, and generally speaking, have upheld its use. Therefore, one will continue to find the term in patent specifications.

Another word is "illustrative," likewise of frequent currency. It is another hedging word. It says that what I am describing is just an example to which I do not wish to be limited, since there are many others that I could also give.

"Plurality." Obviously, if I do not wish to be limited to any given number greater than one, "plurality" is exactly the word to use, and many instances of its use will be found.

"Multiplicity." Where it is plainly evident that a large number of items are called for, "multiplicity" is a good word to use. If, however, the word is used in a claim, a question will arise if the competitor, let us say, gets down to as few as three. As far as I am aware, there has been no judicial determination indicating the lower limit of coverage afforded by this term.

"Embodiment" likewise has a secure place in the patent vocabulary; it is used in the Rules of Practice (17). Inventions are universally recognized to consist, first of all, of a mental concept termed "conception," but until such concept takes physical form, an invention has not been completed. When the mental concept does take physical form—that is, when the invention is reduced to practice—the patentee is likely to refer to this as an "embodiment" of his invention. Frequently "specific embodiment" is used. The inference is that other embodiments are equally likely.

We have already referred to specific examples which are on a par with specific embodiments. Usually in mechanical inventions, we speak of "specific embodiments," whereas, in chemical or metallurgical inventions, we speak of "specific examples."

Metallurgical Patents

If thus far metallurgical patents have not been mentioned specifically, it is because essentially the same rules apply to the patenting of metallurgical processes as apply to any other chemical process. Alloys fit nicely into the category of compositions of matter. However, one or two cases have arisen in the metallurgical art that are peculiar thereto and deserve brief mention.

We may consider the Coolidge patent (2), which represented a great advance in the art of metallurgy of tungsten, since Coolidge had taken the normally brittle element of nature and, by his process, converted it into a metal having the seemingly entirely new property of great ductility. In this form it was capable of being drawn into wire, and the wire could then be used for filaments in incandescent electric lamps. Unfortunately for the patentee, when the patent was sued upon, it was brought out that pure tungsten, instead of being brittle, is highly ductile. This vitiated the product claims and left Coolidge without protection on ductile tungsten. The case is still cited for the proposition that one cannot patent a product of nature.

In drawing patent specifications to cover alloys, the question of specific proportions becomes extremely important. If a broad range of proportions is sought to be claimed, there must be adequate specific examples upon which to base such broad claims. Furthermore, such claims involve to a high degree, the distinction between the words "composed of" or "consisting of," which are excluding in character, and "comprising," which is nonexcluding in character.

Thus the Gray patent (6) covered the discovery that a minute amount of indium, when incorporated with silver, gave protection against tarnish. The inventor attempted a claim reading:

"A tarnish-resisting alloy . . . *comprising* silver and indium with the silver content predominating and the indium in sufficient quantity to give protection to the alloy or mixture, against tarnish."

It was held, however, that this claim was too broad, because it would cover alloys containing quantities of other elements, and these might result in an alloy which was not tarnish-resisting. The words "composed of" and "comprising" figure prominently in most all compositions of matter claims and also in most process claims. Many times an examiner will act, as in the Gray case, by holding that there is nothing in the specification to show that the invention is broad enough to warrant the use of "comprising." This is another instance where pains must be taken to make the disclosure of the specification sufficiently broad, if a broader terminology of the claims is sought.

Finally, no mention of the subject of alloys would be complete without mentioning the Marsh Nichrome case (*13*). Marsh admitted that the alloy itself was old; his discovery was for a new use—namely, as an electric heating element. Therefore he was allowed to claim an electric resistance element formed of a metal alloy consisting of nickel and chromium. Present practice no longer permits claiming a new use for old materials in this fashion; rather a new use must be claimed, if at all, as a process in which the new use of the known material is recited, all as set forth in Section 100(b) of the Patent Law.

Interpretation

Determining the scope of a patent is usually a fairly technical proceeding and should be reserved for the specialist. The final answer must, of course, rest with the court of last resort. While the Supreme Court of the United States has the last say in the matter, it is rather seldom that a patent reaches the high court for adjudication. Both patent lawyers and judges find the determination of the question a difficult one, and even the examiners in the Patent Office, while the patent is pending, do not always have an easy time in deciding what is or is not patentable. Of course, the fact that lawyers often differ in their estimates of the scope of a patent makes for law suits. In this respect law suits on patents do not differ from those involving other branches of law.

People engaged in technical library work frequently are given the task of finding an anticipation of a patent, or as it is sometimes expressed, "finding the pertinent prior art." This will be useful in estimating the validity of a patent, either when a license is sought or asked or when the patent is sued upon. Since the examiners in the Patent Office have only limited time for searching before issuing the patent, it is unwise to rely upon their search as being conclusive of validity of the claims, although considering the amount of time available to the Patent Office, the examiners' searches are, in most instances, of high quality.

How does one know when an anticipation has been found—that is, a prior patent or publication which will affect the validity of the patent? The simplest test: Does each of the claims of the patent read upon the reference? If it does, one may assume that it is such an anticipation. Perhaps the language does not read literally upon the reference, but nevertheless it becomes apparent that at least a portion and perhaps much of what the patentee and the Patent Office thought was new was not new at that time. The question then becomes: Is the residue of novelty sufficient to sustain the patent? One can only touch upon this phase of the matter, since no one to date has evolved a satisfactory affirmative definition of invention or what in Section 103 is now termed "nonobvious subject matter," although many yardsticks have been proposed which state what is "not invention."

Having now some notion of how a patent specification should be drawn, let us apply our knowledge to an actual patent. The patent I wish to discuss—Jones, Kennedy, and Rotermund (9)—issued on June 9, 1936, originally to Union Carbide & Carbon Corp. An examination of the reports reveals that this patent has the distinction of being the last one to have been sustained by the U. S. Supreme Court. To be sure, since the Union Carbide case, the high court has taken a number of patent cases for consideration, but save for the so-called "A&P case" (1), in which the patent was held void, none required consideration of the issues of validity and infringment.

The Jones patent (9) expired June 9, 1953, so that anything said about it will be, to use a favorite legal word, "moot." However, a discussion of the patent and the rulings of the courts thereon, will serve not only to illustrate the matters that have been previously discussed, but some that we have not as yet touched upon.

In customary fashion, the patent starts off with a statement as to what the invention relates—that is: "This invention relates to electric welding." Following this, the patentees describe sufficient of the prior art to afford a background for the proper understanding of the invention to come. After discussing several of the prior-art practices, the closest prior-art method is described—the so-called "protective flux" method. The flux, described as consisting of natural clay, is placed over the surfaces to be welded, and the arc is struck under this powdered flux. The principal drawbacks attending this operation involving this flux, are recited—that is, the weld is porous, and the arc projects a continuous cloud of material into the atmosphere, necessitating that the welders wear gas masks.

We now come to another important part of the specification—namely, recitation of the objects of the invention. They are all stated as relating to a method of welding in which the itemized drawbacks are overcome. And then we find the usual stereotyped expression: "Other objects of the invention will become apparent as the description of our invention proceeds."

This stereotyped expression was perhaps fortunate, inasmuch as a new fluxing composition, which turned out to be the only feature of the invention the court considered patentable, was not included in the stated objects of the invention. Even more fortunate was the statement: "The composition of the welding medium is of the utmost importance."

Then there is a recital of the compounds used in the new flux, as follows: "We prefer to use silicates of the alkaline earth metals, such as calcium silicate, and we also prefer to add to these silicates minor proportions of alumina and of a substance adapted to lower the melting point—for example, a halide salt." The patentees continue: "We have used calcium silicate and silicate of sodium, barium, iron, manganese, cobalt, magnesium, nickel, and aluminum"—nine in all. This language, in view of the litigation involving the patent, proved to be of the greatest significance.

Finally, the specification, after giving detailed information on how to practice the invention, concludes with this language:

"This application is a continuation-in-part of our prior applications serial Nos. 657, 836 and 705, 892, respectively, filed Feb. 21, 1933, and Jan. 9, 1934."

We may pause briefly to consider the significance of this statement. It illustrates a practice frequently indulged in and stems from the fact that the Patent Office has a salutary rule that once an application is filed, it cannot thereafter be amended to contain what is termed "new matter" —that is, matter that is disclosed neither in the drawing nor the specification when the application was filed. Frequently, however, after an application has been pending for awhile, new aspects of the invention come to light, and many times the inventor finds that what has been stated as to the theory of the invention has been erroneous. Or the citation of a heretofore unknown reference has presented him with a dilemma. The remedy under these circumstances is to commence all over again—that is, to file a brand new application in which new matter will be added and the old matter corrected, if necessary. Having done this, the old application will become abandoned, but in accordance with the statutes (Section 120), the new application, called a "Continuation-in-Part," will be entitled to the benefit of the filing date of the original case for all material that is common to the two cases and carried forward to the new case. In the patent under discussion, not one but two such applications were filed.

The court opinions bring out the reasons for two continuations-in-part. If possible, the patentees wished to establish and claim a new method of welding, rather than having to rely upon claims to the fluxing compounds, knowing full well the difficulty of covering all such compounds that would be likely to work.

Four court decisions (*5, 11, 12, 18*) passed upon this patent.

Just what did the trial court do with the patent? It held first of all, that an invention of great merit had been made and recognized that the patentees had been able, by the use of their improved fluxing composition, to conduct their electric arc-welding operation without glare, with no open arc, no splatter, and very little, if any, smoke. Further the performance achievement was held to be far superior to what had gone before. The courts were unanimous in recognizing the invention as a meritorious one. In spite of this recognition, however, the trial court, while upholding some of the product claims, held invalid the process claims on several grounds. We may consider those relating to claim 11:

"A process of electric welding which comprises the step of forming a conductive high resistance melt, containing a major proportion of alkaline earth metal silicate and substantially free from uncombined iron oxide, on a metal part to be fused; and passing an electric current through a circuit comprising said melt and said metal part."

One of the grounds of invalidity was that the process claims recited the same operational steps as the prior art, though conceding at one point that the patentees had discovered a new process.

Another ground, the court found, was that the process claims had been predicated upon an error. The claims called for having the electric current to be conducted by the molten flux, whereas properly speaking, the current was conducted in part, if not in its entirety, by the action of the electric arc.

Further, the court held the process claims invalid because they were too broad and indefinite and hence did not comply with the then Section 33 of the patent statutes; this is none other than our old friend, Rule 75, which states that the applicant shall particularly point out and distinctly claim the part, improvement, or combination which he claims as his invention or discovery. In considering these last two grounds, we may recall that it has been frequently held that a patentee is not to be charged with a knowledge of the correct theory or even understanding of his invention (21). Yet we see that, as a practical matter, a misunderstanding of the theory was one reason which caused the court to strike down the method claims.

However, the trial court held some of the product claims valid and others invalid; thus, claims 24 and 26, which called for either metallic silicates or just silicates broadly, were declared invalid on the ground that individual silicates were old for the purpose. Claims 18, 20, 22, and 23 were upheld. These called for alkaline earth silicates or calcium silicate.

We now find that the court has narrowed the patent to four product claims, and, as to these, the defendant can argue that he does not come within their terms. The reason: The claims now left all call for the welding composition to have a major proportion of alkaline earth silicate, whereas the defendant's composition does not contain a major proportion of alkaline earth metal silicate, but rather a major proportion of manganese silicate. Manganese, however, is not classified as an alkaline earth metal, so that now the only thing that can save plaintiff's case is to have the court hold that manganese silicate is the equivalent of calcium or other alkaline earth silicates. Fortunately for the plaintiff, the court so held and called attention to a passage in the patent specification, already quoted, in which the patentees say they have used, among others, manganese silicate. So we arrive at the final result so far as the trial court is concerned: The patent has been upheld as to certain of the product claims, and the defendant has been adjudged as infringing.

Before following the fortunes of the parties litigant in the Court of Appeals, let us consider a rather fundamental question which concerns patents and their interpretation. In arriving at its decision, the court considered the two fundamental questions which are present in every patent suit, unless one of them is conceded or waived by the defendant. These are: (1) Is the patent valid, and (2) if valid, is it infringed? As you will infer, the court decided both questions in the affirmative, although some claims were held to be invalid.

Now the reason why patents are usually taken out with more than one claim will be apparent. In the patent in suit, there were 29 claims, and at the time the patent was being solicited, there was no way by which the then applicants could have foreseen that of these 29 claims, four would have been held valid, while most of the rest would have been held invalid. Some of the claims had not been sued upon.

The Case on Appeal

Both parties appealed to the Court of Appeals of the Seventh Circuit. Here the court sustained the judgment of the trial court as to the product claims already held valid, and, in addition, reversed the trial court by holding claims 1 to 9, 11 to 18, 20, 22 to 24, 26, and 27 valid, though not infringed, and to this extent, reversed the decision of the trial court. Presumably the claims not mentioned had not been sued upon.

However, the defendant still had another remedy at its disposal, although let it be understood, one that is not too often realized. The defendant sought a writ of *certiorari* to the Supreme Court, and this was granted. It must be understood that a review on writ of *certiorari* is not granted as a matter of right, but only in the sound judicial discretion of the Supreme Court. We are not told the exact grounds upon which the writ was granted; most frequently it is granted where there are conflicting decisions on the same patent in differing Circuit Courts of Appeal. This situation did not obtain here. The Supreme Court also grants the writ in cases involving a matter of great public concern. Presumably this court was impressed by the complete reversal by the Court of Appeals of the trial court's holding of invalidity of the process claims. After a considered opinion, the Supreme Court reversed the Court of Appeals to the extent that, in effect, it reinstated the decision of the trial court, thus leaving the parties as they were prior to the appeal to the Circuit Court of Appeals of the 7th Circuit.

One might well think that having survived the perils of the U. S. Supreme Court, the plaintiff's worries would be over, and it could go on its way safe in the assurance that the last word had been spoken by the highest court in the land. However, the plaintiff had not reckoned with the possibility that the court might grant a rehearing on the issue of infringement. This the Supreme Court did (*18*). However, the majority of the court held for the patent, and refused to overturn its previous decision. The Supreme Court, as were the courts below it, was impressed with the fact that the defendant had not engaged in any substantial independent research but had merely followed the teaching of the patent itself.

The majority opinion was accompanied by a strong dissenting opinion written by Mr. Justice Black, with whom Mr. Justice Douglas concurred. The dissenting opinion alluded to the doctrine that that which is disclosed but not claimed, is dedicated to the public. Further, the dissenting opinion held that it was an injustice to the public to go contrary to this accepted rule of law, and pointed out that Congress had already provided a remedy for the patentees under these circumstances. This was that a reissue should have been sought, in which the question of adding new claims— that is, a claim reciting manganese silicate—would have been threshed out before the Patent Office where it properly belonged.

After the second Supreme Court decision, the assignee of the patent now proceeded to act under Sections RS 4917 of the old Patent Act and filed what is known as a "disclaimer," and we find this disclaimer now printed as a part of the patent (*9*), as follows:

"Lloyd Theodore Jones, Harry Edward Kennedy and Maynard Arthur Rotermund . . . hereby enter this disclaimer to claims 1 to 17 inclusive, and claims 24, 26, 27, 28 and 29 of said specification."

The subject of disclaimers underwent considerable revision in the new Patent Code of 1952. The effect of this was to reduce greatly its effect on the patent as a whole.. More than a century ago, if one or more claims of a patent was held invalid, the entire patent was likewise invalid. To relieve the harshness of this rule, the original disclaimer statute provided, in 1837, that the patentee could still sue on the other claims, provided a so-called disclaimer were filed and, provided further, that the patentee had not unreasonably neglected or delayed to enter the disclaimer. This language was fruitful of much controversy and has now been done away with. The only penalty now for not filing a disclaimer is that the patentee can no longer recover costs of the suit. The disclaimer in the patent under discussion was filed in 1949, before the new patent statute became effective Jan. 1, 1953.

Markush Claims

The Court of Appeals of the Seventh Circuit, in effect, construed certain of the product claims of the Jones patent as including the nine metallic silicates that the patentee found to be operative. However, neither the trial court nor the Supreme Court would follow this line of reasoning, although they did hold that manganese silicate was to be regarded as an equivalent of the claimed calcium silicate.

It is interesting to speculate whether the patentees, while their application was pending, could have claimed an artificial grouping of the silicates in more definite language that would have embraced all of the species—for example, "a fluxing composition for electric welding . . . selected from the group consisting of alkali metal silicates, alkaline earth silicates, and manganese silicates." While it is not entirely clear that the Patent Office would have allowed such a claim at the time in addition to the natural genus (although at the present there is good precedent for it), it does illustrate the practice of creating an artificial grouping which is known as a "Markush" claim. If the patentees could have had such a claim, it may be assumed that they would have avoided the hurdle of having to have manganese silicates adjudged the equivalent of the alkaline earth silicates, which was only surmounted with the greatest difficulty.

This artificial genus received the name "Markush" when a chemist-inventor named Markush, in 1925, prevailed upon the Assistant Commissioner of Patents to approve as to form, a claim which read "material selected from a group consisting of aniline, and homologues of aniline." This decision, which seemed to be dictated by necessity, established a precedent which has been extensively followed to this day and has conferred a sort of immortality upon the name of Markush.

Reissues

In their dissenting opinion (*18*), Mr. Justices Black and Douglas said that if the Jones patent (*9*) did not directly claim the invention, a reissue should have been applied for. Here again, we have a subject that can only be touched upon most lightly.

The subject of reissues is dealt with in Sections 251 and 252 of the Patent Act. Essentially, they provide for correction of errors in the patent grant which arise by reason of a defective specification or drawing or by reason of the patentee claiming more or less than he had a right to claim in the patent, provided the error arose without any deceptive intention. The procedure requires an offer to surrender the original patent and an application for the reissue thereof. If the reissue is granted, the surrender of the original patent becomes effective, and the reissue is granted for the unexpired term of the original patent. The reissued patent may not, as a rule, be enforced against one who, prior to the grant of the reissue, has acquired what is termed an "intervening right."

For example, let us suppose that the decision of the trial court had been entirely adverse to the plaintiff and that the patentees of the welding patent (*9*) had then applied for a reissue and had received it. It is then clear from the language from Section 252 of the statute that they could not have held the defendant under the new claims acquired by the reissue for continuing to do what it had done prior to the grant of the reissue patent.

We have dwelt at some length upon the process of obtaining patents and what can happen to a patent once granted, with the thought that such knowledge will be helpful in understanding them. Essentially, the grant of a patent is an adversary process, in which the patent examiner, representing the public, and the patent lawyer, representing the inventor, finally arrive at some middle ground which, it is calculated, will best define the scope of the rights of the inventor without undue deprivation on the part of the public.

Based on a paper presented before Columbia University School of Library Service, Institute on Patents as a Source of Information, June 30, 1960, revised April 1964.

Literature Cited[1]

(1) Atlantic & Pacific Tea Co., 87 U.S.P.Q. 303.
(2) Coolidge, U.S. Patent **1,082,933** (1913).
(3) Dosselman and Neymann, In re, C.D. 379 (1911).
(4) Ellis, In re, C.D. 374 (1911).
(5) Graver Tank Mfg. Co. v. Linde Air Products, 80 U.S.P.Q. 451.
(6) Gray, U.S. Patent **1,847,941** (March 1, 1932).
(7) Grimme, In re, 124 U.S.P.Q. 499.
(8) Isenstead v. Watson, 115 U.S.P.Q. 408 (1957).
(9) Jones, Kennedy, & Rotermund, U. S. Patent **2,043,960** (June 9, 1936).
(10) Krimmel, In re, 130 U.S.P.Q. 215 (1961).
(11) Linde Air Products v. Graver Tank Mfg. Co. (Trial Court decision), 75 U.S.P.Q. 231.
(12) Linde Air Products v. Graver Tank Mfg. Co. (7th Cir.) 77 U.S.P.Q. 207.
(13) Marsh, U.S. Patent **811,859** (Feb. 4, 1906).
(14) Nelson and Shabika, In re, 126 U.S.P.Q. 242 (1960).
(15) Tolkmith (P.O. Bd. App.), 102 U.S.P.Q. 464 (1954).
(16) 159 U.S.C. 564 (1895).
(17) U.S. Patent Office, "Rules of Practice in Patent Cases," 4th ed., Rules 71, U.S. Government Printing Office, Washington 25, D.C. (1960).
(18) U.S. Supreme Court, 85 U.S.P.Q. 328.
(19) Warp v. Warp, 48 U.S.P.Q. 505 (1941).
(20) Welsbach, U.S. Patent **830,017** (1906).
(21) Westinghouse Electric Co. v. Montgomery, 153 Fed. 890,901.

---

[1] Legal abbreviations are defined on page viii.

RECEIVED January 17, 1964.

8

Patentability of Homologs, Isomers, and Other Analogs

DEAN LAURENCE

Laurence and Laurence, 753 Warner Building, Washington, D. C.

In determining the patentability of novel homologs, our concern is the "obviousness" statute in our patent law, 35 U.S.C. 103. The phrases discussed are ". . . the subject matter sought to be patented . . ." and ". . . the subject matter as a whole . . ." These phrases do not mean the same thing. "Obviousness" under Section 103 is a problem of patent law answerable only on the evidence presented as to differences in properties as between a known compound and a claimed compound. The patentability of a compound does not depend on dissimilarity in formulas but on dissimilarities of the tangible embodiments of the two formulas. We have "homologous" cases in the law as well as in chemistry, and it is only upon a study of such cases that a reasonable prediction as to patentability can be made.

Patent law is wholly a creature of statutes. Disputed cases in this field can be decided only in concord with what a statute says. It is only rarely a statute is sufficiently definite or applicable to determine by its precise terms a right result in any particular case. The judicial process of interpreting statutes is a matter of comparing the facts in a case in dispute with the facts in previously decided cases and then making a decision as to which of the usual two lines of precedents appears applicable.

At the outset let it be understood, I shall not here endeavor to define precisely the subject terms with which I am supposed to be dealing. I think the answer immaterial to determinations of patentability. The result of a patentability holding arising only out of a failure to find homology or isomerism is fast disappearing. The issue of patentability of a compound today involves the threshold question as to whether the prior art describes in any way a compound so closely related in theory of its structure as to raise a presumption of obviousness in a compound sought to be patented.

Intelligently to discuss the assigned subject of the patentability of a compound closely structurally related to a prior art compound, I am compelled to give you two hard bits upon which to chew. The first is known popularly as the "obviousness" statute in our patent law—that is, Section 103 of Title 35 of the United States Code. The second, to which I shall advert later and more briefly, is the necessity for understanding the value of consistency and generality, or the "seamless web," of the whole body of patent law.

Obviousness Statute

I shall ask you to look at the first sentence of Section 103, because this is the section under which patentability of the subject matter here involved is determined.

"A patent may not be obtained though the invention is not identically disclosed or described as set forth in section 102 of this title, if the differences between the subject matter sought to be patented and the prior art are such that the subject matter as a whole would have been obvious at the time the invention was made to a person having ordinary skill in the art to which said subject matter pertains."

I direct your attention to the two phrases thereof as follows: ". . . the subject matter sought to be patented . . ." and ". . . the subject matter as a whole . . . ." These phrases do not mean the same thing, and upon their distinction is predicated the patentability of subject matter categorized as homologs, isomers, and analogs. The phrase "the subject matter sought to be patented" means the mental concept defined by the claim of a patent application. It is the name or theoretical formula of a compound. The phrase, "the subject matter as a whole," means the tangible embodiment of that mental concept including its inherent applied use characteristics. It is the physical thing and its attendant properties. While I am unaware that any court has thus precisely so defined these phrases, I submit the foregoing is a succinct summation of the discursive reasoning advanced in the opinions in such decisions as that by the United States Court of Customs and Patent Appeals (2).

Were we compelled to look solely to the differences between the naked structural concept of the subject matter sought to be patented and the naked structural concept of the prior art, the type of chemicals here discussed would simply not be patentable. For example, the broad mental concept of a next-adjacent homolog of every known organic compound capable of having such relationship is known to every chemist. But this does not mean such homologs may not be patented *per se*. It does mean the patentability of such compounds is dependent upon the evidence presented to an examiner or a judge as to the "subject matter as a whole," which includes the inherent use properties of the tangible embodiment of a claimed name or formula concept. I refer to the wonderful paragraph in the opinion by Judge Rich (In re Papesch) as follows:

"From the standpoint of patent law, a compound and all of its properties are inseparable; they are one and the same thing. The graphic formulae, the chemical nomenclature, the systems of classification and study such as the concepts of homology, isomerism, etc., are mere symbols by which compounds can be identified, classified, and compared. But a formula is not a compound and while it may serve in a claim to *identify* what is being patented, as the metes and bounds of a deed identify a plot of land, the *thing* that is patented is not the formula but the compound identified by it. And the patentability of the thing does not depend on the similarity of its formula to that of another compound but of the similarity of the former compound to the latter. There is no basis in law for ignoring any property in making such a comparison. An assumed similarity based on a comparison of formulae must give way to evidence that the assumption is erroneous."

Two Different "Somethings"

For sometime the Patent Office entertained the view, first distinctly expressed in a dissenting opinion (*1*) to the effect, "how can something which is obvious, logically become something which is unobvious upon a showing of comparative superiority?" While such question appears logically impeccable in requiring a negative answer, inspection and elaboration of the question makes it clear our old friend—namely, the logical fallacy of the "undistributed middle" in syllogisms—is inherent in the question. The expression "something" was utilized in the dissent in two different ways to mean the same thing, while, in fact, Section 103 refers to two different "somethings."

I return again to the phrases "subject matter sought to be patented" and "subject matter as a whole." In the instance of a novel compound, a patent claim defines a mental concept, a theoretical structure, the "something" by which its physical embodiment can be made recognizable in name or picture, and this is the "subject matter sought to be patented." This claimed structural concept, novel but sufficiently closely art-related as to raise a presumption of obviousness, is the first "something" of the dissent. And, absent a second "something" in the statute, Section 103, that would be the end of the matter. Patentability would be decided upon a mere subjective or visceral determination of the "something" claimed being so closely related to "something" known as to be obvious.

However, the statute fortunately does direct our attention to a second "something"—that is, "the subject matter as a whole"—and affords a more objective basis for determining "obviousness." Thus, the presumption of obviousness apparent upon inspection and comparison of the names or formulas of the claim and of the art is only what lawyers call a "rebuttable" presumption. If the compared names or theoretical structural drawings are so similar as to cause reasonable judges to conclude temporarily the claim covers subject matter differing only obviously from the names or structural drawings of the art, then the physical embodiments of the names must be compared to dissipate or confirm the temporary conclusion. We must compare what useful attributes the physical embodiments expectedly have in common and, unexpectedly, not in common, and on such facts conclude as a matter of law which way the decision ought go.

Section 103 does not mention such subject matter as homologs, etc. In fact, it contains no clause specific to chemical things. Yet the issue presented by cases in the area here considered must be decided by recourse to its language. The only way open to patent attorneys and patent-minded chemists of getting at an answer on the patentability as to any closely art-related compound is by comparison of the facts in a disputed case with those in the decided cases in this particular area. We have "homologous" cases in the law as well as in chemistry.

That the determination of patentability under Section 103 is not a chemical problem is an observation for which I do find support in the opinion of the C.C.P.A. (2), as follows:

"That problem of 'obviousness' under section 103 in determining the patentability of new and useful chemical compounds, or, as it is sometimes called, the problem of 'chemical obviousness,' is not really a problem in chemistry or pharmacology or in any other related field of science such as biology, biochemistry, pharmacodynamics, ecology, or others yet to be conceived. It is a problem of *patent law*."

The court summarized the whole matter in this way:

"What this comes down to, in final analysis, is a rather simple proposition: If that which appears, at first blush, to be obvious though new is shown by evidence *not* to be obvious, then the evidence prevails over surmise or unsupported contention and a rejection based on obviousness must fall."

Evidence Can Rebut Obviousness

There will be no "problem" of patent law, or, perhaps I ought say, no problem for the chemist, the patent attorney and Patent Office, or courts, unless the chemist and those allied with him in evaluating the physical compounds he makes properly perform their tasks in accumulating the evidence whereby a problem can be presented. What must be the nature of such evidence?

Let us go to the subject of next-adjacent homologs and assume "methanol" known and "ethanol" the novel subject matter sought to be patented. We face an immediate question as to what "manner of using" the ethanol was asserted in the patent specification in conformity with the requirement of Section 112 of the statutes for such disclosure, and we will here make an assumption that the disclosed use is as a solvent in formulating an elixir. We will make the further assumption that the art taught methanol to be a solvent for several organic compounds.

The Patent Office naturally says the claim to ethanol is rejected as being for a composition obvious to one skilled in the art having knowledge of methanol and the fact that both the asserted elixir-solvent use and the art-solvent use are the same functionally.

The evidence necessary to rebut this presumption will reside in the pharmacological finding that methanol in an elixir generally kills people and cannot be used, while ethanol does not and may serendiptiously enhance at least temporarily their feeling of well-being. Thus, the obvious mere "CH_2" difference between the naked structural concept of the "subject matter sought to be patented" (CH_3CH_2OH) and the naked structural concept of the prior art methanol (CH_3OH) was not such that the "subject matter as a whole," which includes the unexpected use property of the embodiment of ethanol (CH_3CH_2OH), turned out to be something one skilled in the art could have foreseen. The assumed similarity of the substances, based on a comparison of their formulas, must give way to the evidence the assumption is erroneous. Ethanol would be patentable.

Another situation can be illustrated hypothetically by considering the normal- and iso- position isomers of propyl alcohol. Assume the normal alcohols are known, but branched-chain alcohols are novel although branched-chain alkanes are known. A claim to isopropyl alcohol will be rejected as directed to a composition obvious because of its close relation to normal propyl alcohol in view of the art showing branched-chain alkanes. How could this be overcome? We must call upon the resourcefulness of the allied sciences in an effort to find some use property of the embodiment of the subject matter sought to be patented not inhering to the *n*-propyl. Under the case law apparently prevailing, such use property must be asserted in the application as filed, else an applicant will not be permitted to present comparative showing. In some respects this does not make a great deal of sense to me, but that is the way it is presently, and one must therefore determine in advance of filing the nature of how a case will be argued *vis-a-vis* the art. This means an applicant must make a thorough search and know the art. It also means the specification ought to state clearly what the structural modification is that constitutes the difference between what is claimed and what is old.

Among chemists in the medicinal field, one would avail himself of the services of a pharmacologist in an endeavor to differentiate the properties of the isomers. Suppose you are not in the drug field—what can you do? The answer is that you will be associated with workers in some utilitarian field, and it is to them you must turn. Perhaps the isocompound will serve to preserve eggs for 12 months at temperatures up to 100° F., while the normal isomer causes leaks in the shells in few days at even room temperatures. This is not silly, and it is the sort of evidence needed to rebut the presumption of obviousness arising out of the close structural similarity.

I suppose something ought be said about compounds whose sole known utility is as intermediates or starting materials for making derivatives. I have not yet gotten the courts to agree with me, but I think the inherent chemical property of being useful to make a novel derivative which has unobvious properties is just as good evidence on the patentability of the starting material as any of its other properties and ought to make for patentability.

I digress to comment that some chemists incline to the view that novelty ought be the ultimate criterion of patentability and query all this rubbish about having to prove that a novel compound they have originated is unobvious. They complain they have made one invention and decry having to make a "second" invention to make patentable the first. To this there are two answers. First, the statute law requires it. But, says the chemist, in the memorable words of Dickens, ". . . 'if the law says that, it is an ass' . . . and it ought be changed." So I will give you the second reason which is grounded in sound public policy. In any field there is a certain amount of knowledge of concepts and physical things—published, unpatented, and in our general fund. Any person is entitled to draw upon this knowledge as needed or desired. This is the black area of unpatentable subject matter under Section 102 of the patent statutes. No one can reasonably quarrel with the proposition that what is not novel is not patentable. At the other end of the spectrum is the white area of concepts having such far-out novelty that patentability is unquestionable, and no right-minded examiner or judge would ask for evidence proving their physical embodiments unobvious, because there is nothing reasonably suggestible with which such embodiments could be compared.

In between is the gray area of novel things productive of disputes. Is the subject matter so darkly tinged as to be substantially black and unpatentable or light gray enough to be patentable? Why consider this at all? Why not use mere novelty as the determining factor? The difficulty is that this would run afoul of the Constitution which wisely says a discovery must "promote" the progress of science and the useful arts. We need this gray area. Let me illustrate. Suppose an expired patent described reacting *A* with *B* to make *C* at a temperature between 50° and 100° C. with a yield of 60%. Another chemist runs the reaction at 105° F. and gets a yield of 65%. He has novelty, but would you contend he ought to be allowed a patent on such finding on the basis of his mere novelty? I think you would certainly adopt the view, if you wanted to use the published process, that you were entitled, as an ordinary chemist, to fool around with closely related temperatures to see whether you could improve the yield a little. Patents granted on mere novelty would be bad for the general public because they fail to promote the art. The same consideration applies to compounds. Making another novel compound which differs only as by close homology, isomerism, sulfur for oxygen, or a double-bond shift is an exercise in manipulative chemical procedures unless such obvious difference produces a physical thing of unexpected usefulness.

Patentability Cannot Be Determined by Rules

There can be no rule susceptible of mathematical application to any particular compound for determining patentability. Whenever you are confronted with the question: Study the related art from a structural standpoint; set down clearly the structural difference; study what the art says as to the effect of such difference in the structures of the other related compounds where such structural difference already exists; examine all possible facets of the effects of the structural difference in your new compound compared with the structurally related art; and, above all, chew over and appreciate the significance of every fact you uncover. Then study the case law and make your guess as to the result of your patentability study.

Earlier I suggested regard for the seamless web of the body of law. Patent law is but a small fraction of our total law, and the cases concerned with chemical patent law are but a small part of this fraction. There is no statute which specifically concerns itself with problems of the patentability or validity of chemical subject matter. Cases in our field must be in harmony with the sound doctrines established in the field of things mechanical long before the chemical art achieved importance. I think chemists turned examiners, patent agents, and lawyers all give far too little attention to the opinions of the courts in cases involving other subject matter. The rule, "the addition of an element to a patented combination without a distinct change in function does not avoid infringement," is an example of this. Why should the addition of a methyl group on one ring of a chicken-wire compound either make for patentability or avoid infringement unless the physical embodiment thereupon acquires an unexpected use property? We have one substantive statute, Section 103, for determining patentability of nonidentical things; it applies to all statutory classes of things; and, so do the general rules of the cases decided under this statute.

Literature Cited[1]

(1) Brody, Ex parte, 122 U.S.P.Q. 611.
(2) Papesch, In re, 137 U.S.P.Q. 43.

---

[1] Legal abbreviations are defined on page viii.

RECEIVED October 10, 1963.

9

Patent Protection Available on New Uses for Old Chemicals

S. BRANCH WALKER

American Cyanamid Co., Stamford, Conn.

> Chemicals are classified as compositions of matter from a patent statute viewpoint. A claim on a composition protects the inventor without regard to the use of the product. Where a new use is found for an old chemical, if the new use incorporates a change in the physical form or packaging, sometimes product claims can be obtained. More commonly, the new discovery must be claimed as a method of using. Examples are given to illustrate the type of wording which has met acceptance in the Patent Office. No matter how worded, the legal requirements for patentability, particularly unobviousness, must be met.

Basically a patent can issue on a new chemical, and the patent covers the chemical itself—independent of the use to which the chemical may be put. Similarly, if the chemical is old, a new use for that old chemical does not make the same chemical, which is old, appear as new. Assuming the new use is invention, the problem is to obtain patent protection on the new invention, which for the purposes of this article is a new use. Hence, that aspect which is new must be pointed out and claimed.

A patent is a creature of statute. The inventor must describe and claim his invention in terms of the statutes to obtain protection. Statutes are the basis for patent law, and to utilize the provisions of the patent laws, descriptions and claims covering inventions must be conformed to fit the law. Adapting the statutes to fit a set of facts is rather rare, although it has been done. For example, after World War II, the life of certain patents was extended. In Radio Position Finding Corp. v. Bendix Corp. (37), a special act of Congress had waived the one-year limit on public use for a particular Blair application on radar. Military security had prevented timely filing. The court found such special law to be proper.

In chemistry, the same results come from the same procedure. If a different result is obtained, it is because of some variation in the procedure. It may be difficult to locate the variation, but it is there—it can be found. Also, a chemical reaction will proceed the same in any country and at any time. Our understanding of the laws of chemistry may change, but the laws of chemistry themselves do not change.

The laws of men and governments are not so rigid. The laws themselves can be changed, and the understanding, interpretation, and application of statute laws can change. Thus, any present law pertaining to patents can be reinterpreted, or the law itself can be changed. A new patent act became effective Jan. 1, 1953, in the United States.

Section 101 of the U.S. Patent Law (*51*) provides for patents in the following categories:

1. Processes
2. Machines
3. Manufactures
4. Compositions of Matter

Plant patents and design patents are separate statutory classes but are not here pertinent.

In considering new uses for old chemicals, each category should be considered to determine if useful protection can be obtained in that category.

Composition of Matter Claims

Chemicals as such are not listed as a patentable category. Actually, the term "chemicals" has been used with different meanings by different people at different times and is not susceptible of a really tight definition. Frequently "chemicals" means compounds. Compounds also are not a statutory class but fit into the broader class of compositions of matter.

A composition of matter not only has a chemical structure for one or more components but also has physical characteristics which may be important in securing patent protection.

In re Papesch (*33*) holds (italics added):

". . . *a formula is not a compound* and while it may serve in a claim to identify what is being patented, as the metes and bounds of a deed identify a plot of land, *the thing that is patented is not the formula but the compound identified by it.* And the patentability of the thing does not depend on the similarity of its formula to that of another compound but of the similarity of the former compound to the latter. There is no basis in law for ignoring any property in making such a comparison. An assumed similarity based on a comparison of formulae must give way to evidence that the assumption is erroneous."

When patent protection is sought on a new use of a chemical, a first and important question is whether that new use requires a particular degree of purity, crystalline phase, admixture with diluents, location, or storage facilities. If any of these or any other physical attribute is both novel and essential to the new use, a product claim limited to the new attribute should be considered.

An example of invention in which a new crystalline form of an old compound resulted in product claims is a Pfeiffer patent (*34*). Claim 1 thereof reads:

"A solvent stable, tinctorially strong unsubstituted copper phthalocyanine pigment in the 'R' form, the particles of which are characterized by (1) having an average size of less than two microns, (2) being crystalline in structure, (3) which exposed to X-rays in an X-ray diffraction apparatus having a diffraction pattern with the line of maximum intensity corresponding to an interplanar spacing of 11.6 A., the second most intense line corresponding to a spacing of 9.66 A., and a third line at 3.14 A., (4) when subject to infra-red radiation having characteristic absorption maxima at 11.49, 12.91 and 13.74 microns, and (5) yielding the conventional red-shade form pigment when subjected to acid pasting."

This Pfeiffer patent is doubly interesting because of a patent to Wiswall (*59*); claim 8 reads:

"A new and improved, solvent stable, tinctorially strong, halogen-free, sulfuric acid stable, copper phthalocyanine pigment in highly particulate form, the particles whereof are characterized in that (1) they have an average size of less than two microns; (2) they are crystalline in structure; (3) when exposed to X-rays in an X-ray diffraction apparatus they produce an X-ray diffraction pattern in which the line of maximum intensity corresponds to an interplanar spacing of about 12.7 A., in which the second most intense line corresponds to an interplanar spacing of about 9.7 A., and in which the third most intense line corresponds to an interplanar spacing of about 3.75 A.; (4) they retain their average particle size of less than two microns when boiled in xylene for one hour; and (5) they show no substantial changes in tinctorial strength on prolonged storage in contact with crystallizing organic liquids."

This in turn was patentable over the crystalline phase in which copper phthalocyanine was first known to exist. Hence, two separate patents issued on different physical forms of an old compound because that old compound in the specific crystalline condition and physical modification recited in the claims of each of these patents expressed invention.

The physical characteristics of a pigment are extremely important. Each of these different crystalline forms was new and useful and not obvious to those skilled in the art. Each had different color characteristics.

Each physical form of an old compound is not necessarily separately patentable because in most instances such form would not represent invention, because such form would be obvious to those skilled in the art.

With most compounds, a different crystalline form or particle size would have no special significance.

Among the classic cases in this group are the aspirin cases where a particular degree of purification resulted in effectively a new product because as so purified the material had medicinal uses not previously known. The claim of the Hoffman patent (23) is:

"As a new article of manufacture the acetyl salicylic acid . . . being when crystallized from dry chloroform in the shape of white glittering needles, easily soluble in benzine, alcohol and glacial acetic acid, difficulty soluble in cold water, being split by hot water into acetic acid and salicylic acid, melting at about 135 degrees centigrade, substantially as hereinbefore described."

The District Court (14) held:

"That the discovery of the patentee was a most valuable one clearly appears. Even a small amount of free salicylic acid injures the stomach; but, if this can be taken out, the acid is not dissolved in the stomach and does not injure it, but is held in bond intact until it reaches the lower digestive tract. While the discoveries of Von Gilm, Kraut, and others were known for many years before 1898, yet no extensive practical use was ever made of them, while the patented product went into immediate use and so continues on a large scale.

"It is true that Kraut produced acetyl salicylic acid in an impure state, having the same formula as the Hoffman product; but it was comparatively useless. Hoffman discovered a method of taking out the impurities which made the product immediately successful to an extraordinary degree. This he did by his discovery of the waterless process of getting rid of the impurities. Unless the patent law is clearly unfavorable, his discovery should be protected. Kraut's product was not beneficially capable of performing the function of a patented article, while Hoffman was the first to make a successful one. He took a comparatively worthless substance and changed it into a valuable one. It was he, and not Kraut or the other famous chemists of the prior art, who gave to the world this valuable remedy."

A group of cases on aspirin, which also sustain patentability, is cited by the court. The patent was the subject of litigation for years.

The decision of the lower court was affirmed by the Court of Appeals, 7th Circuit, in Kuehmsted v. Farbenfabriken of Elberfeld Co. (28). This line of decisions is still being followed and distinguished in appropriate cases. One of the last distinguishing cases is In re Fisher (16). On petition for rehearing, In re Fisher (17), there are two dissents, which cite much pertinent law. The distinction is on the adequacy of disclosure and form of claims. This case also involved a product of nature aspect.

One of the newer cases following the aspirin holding is Ex parte Yale and Bernstein (60). Claim 17, which was allowed, reads:

"The pure, isomer free dihydrochloride of N-(β-diethylaminoethyl) isonicotinamide, melting at about 194-196°C."

The decision in part is:

"[2] We are unable to agree with the examiner that the pure compounds recited in these claims are unpatentable over the brown sludge of Linnell et al. because they possess new and unobvious properties which are not possessed by the brown sludge. We note in this connection that Linnell et al. not only fail to disclose that the sludge is useful for any purpose whatsoever, but the examiner does not deny that it is not useful for appellants' purpose and that the beneficial properties of the pure salts under consideration are both unexpected and unobvious. The factual situation of the present case, therefore, parallels that of Sterling Drug Inc. v. Watson, cited by the appellants. The rejection of the claims 16 and 17 on Linnell et al. will therefore not be sustained."

Claims to a product were not allowed in Ex parte Steelmand and Kelly (46), where the new product was 13.5 times the potency of the prior art preparation, holding in part:

"[1] There appears to be general agreement in the decided cases that a claim for a known substance which differs from the prior art only in degree, as for example in purity, is not patentable. This principle is illustrated by *In re Mertz*, 25 CCPA 1314; 1938 C.D. 728; 497 O.G. 547; 97 F.2d 599; 38 USPQ 143 cited by the Examiner and *In re Crosley et al.*, 34 CCPA 882; 1947 C.D. 216; 600 O.G. 172; 159 F.2d 735; 72 USPQ 499; *In re Michalek*, 34 CCPA 976; 1947 C.D. 310; 602 O.G. 669; 161 F.2d 253; 73 USPQ 385. An exception to this rule has been made where the purified product possesses a utility not shared by the prior art product as exemplified in the well-known aspirin and adrenalin cases."

This review of some cases based on the aspirin cases is not comprehensive. A complete report on Shepard's Citations (45) on this series alone is very long.

A group of cases on the patentability and unpatentability of purified products has been collected by Biesterfeld (3). Alloys present a different aspect of uses. Both the composition and physical characteristics are critical. Biesterfeld (4) has a collection of decisions in this area. Whether there are actually chemical compounds formed or exactly what are the phenomena as regards phases or the lack of them can be quite vague. In the alloy field it has been held that an accidental prior disclosure within the novel alloy range does not necessarily anticipate, because the accidental disclosures are not concerned with the problem; do not suggest a solution; and may not recognize the existence of such an alloy. In re Tanczyn (48) suggests that, with alloys, the particular form in which the alloy is used may be tied in with the composition. Such a physical construction, for instance a bearing, where the alloy had been used for other purposes but not as a bearing, is patentable (1).

An electrical resistance element has both composition and configuration, and the use of a composition in such a configuration has been held patentable (*26*).

There can easily be argument as to whether a claim to this type of invention should be classed as a composition of matter or a manufacture, as the claim has certain of the attributes of each.

The degree of distinction from a prior composition required to be patentable in part varies with the purposes. Where the closest prior art involved a corrosion inhibitor and a new additive to gasoline was to prevent spark plug fouling (*44*), the court held:

"While the law forbids the granting of a patent for such new use on the theory that patents are granted not for intellectual discoveries but for physical embodiments of such discoveries, nevertheless, as is indicated by Judge Learned Hand, when a new purpose is discovered, slight changes in the preexisting device or composition of material may be sufficient to establish patentability, even if a similar difference without a change of use or purpose might not be sufficient."

The question of patentability of a new use for a chemical as a composition of matter is closely related to that discussed in the paper by Dean Laurence (*29*) on homologs, isomers, and other analogs. The general problem of patent protection is the same because the standards are those of the statute—namely, would the subject matter as a whole have been obvious at the time the invention was made to a person having ordinary skill in the art? Thus the new application not only must use language which distinguishes but must also disclose and claim subject matter that as a whole would not have been obvious. In general, both the Patent Office and the courts are inclined to accept a comparatively small distinction as sufficient for patentable distinction where a truly great invention has been made. On the other hand, where the purpose of use is the same and would be obvious to those skilled in the art, a minor change in the character of a composition is not sufficient to establish patentability. Most of the cases to sustain this particular point, which are known to the average patent attorney, are in the confidential files marked "Abandoned Cases," because no patent ever issued thereon.

A composition of matter patent is greatly to be desired because such a patent covers the composition independent of the time and place of manufacture and the time and place of use and also covers the same composition if used for other purposes.

There has been considerable loose language used to indicate that an inventor is "entitled to all of the uses to which his new invention may be put." This statement is technically correct but easily misinterpreted. The key to the interpretation is that a patent is basically a negative monopoly. The patent owner has the right to exclude others from practicing his invention but, strange as it may seem, the issuance of a patent does not carry the converse right of permitting the patent owner to practice his invention. It may be that one party has a generic claim, and another has a specific claim. The party having the specific claim cannot practice his specific claim because of the dominant generic patent and, similarly, the owner of the dominant generic patent cannot practice the specific embodiment because of the second patent. Thus, the owner of the patent on a composition has rights of exclusion independent of the use to which the composition may be put, but others may also have a right to exclude him under other product or process patents from certain uses of his own invention.

Old Chemical Plus a Carrier

A number of patents have issued on an old chemical with a carrier. Gruskin's patent (*20*) has as Claim 1:

"A cell stimulating composition for use in the treatment of infections comprising a water soluble chlorophyllin dissolved in an aqueous carrier."

The claims of this patent have been held valid and infringed (*40*). A patent on a dilute solution of a particular acid to Morehouse and Mayfield (*32*) was held valid and infringed (*11*). Later this decision was modified, and misuse of the patent held because the product sold was a concentrate for later dilution. The dilute solution was the patented product.

Later, after section 271 of 35 U.S. Code (*53*) on contributory infringement became the law, the Appellate Court (*12*) refused to hold contributory infringement, holding section 271(c) on a staple article was dominant over the rest of the section, presumably 271(b), and refused to reopen the case. Later a new attempt to retry was unsuccessful (*13*). Hence, on the point of contributory infringement, the patent was in essence valid, but ineffective.

The reasoning in a later holding (*18*) is apparently inconsistent on what constitutes contributory infringement.

A Widmann patent (*56*) has, as Claim 1:

"A composition of matter in which the sole essential active ingredients consist of a mixture of tyrosine and pyridoxine."

This patent has been the subject of litigation, including consent decrees (*57*).

An interesting type of claim appears in a patent to Ferguson (*15*), wherein Claim 1 reads:

"A composition useful in digitalis therapy comprising digitalis and a protective colloid consisting of a water-soluble cellulose ether effective in administration to afford final and complete fixation by the heart muscle of the active digitalis glycoside while providing for a sufficiently slow rate of absorption and fixation so that the heart responds with maximum efficiency."

Process Claims

The second of the classes is that of "processes," frequently termed "methods." Although a new inventor cannot get a product patent on a "use," in the sense that it is an old product for a specific use—in effect a new label on the bottle—frequently adequate protection can be obtained by method claims covering the process steps of using the old composition for the new purpose.

Another approach is by a patent on a process for making an old composition where the improved method of making is sufficient to give adequate protection because the old methods are not commercially competitive.

Usually the useful process claims relate to a method of doing something with the old chemical to accomplish a new result. The classic definition of a process is that by the Supreme Court in Cochrane v. Deener (8):

"A process is a mode of treatment of certain materials to produce a given result. It is an act, or a series of acts, performed upon the subject matter to be transformed and reduced to a different state or thing. If new and useful, it is just as patentable as is a piece of machinery. In the language of the patent law, it is an art. The machinery pointed out as suitable to perform the process may or may not be new or patentable; whilst the process itself may be altogether new, and produce an entirely new result. The process requires that certain things should be done with certain substances, and in a certain order; but the tools to be used in doing this may be of secondary consequence."

The novelty of a new method can be either in the material being acted on or in the action taken. As expressed by the Board of Appeals (7):

"It is set forth in the record that the action involved here is one that is not entirely understood. It is considered, however, that it is of a chemical nature rather than physical and that the steps employed are classifiable as a chemical process instead of a mechanical method.

"In such case we find no difficulty in concluding that each chemical agent used constitutes a proper subject for a process claim on the ground that 'although the mechanical steps of immersing the metal in the various baths is the same, the chemical action is different in each case due to the different chemical regents,' and this constitutes a different process.

"The mere steps of manipulation in all chemical processes are relatively few and simple, such as mixing, heating, filtering and distilling, and are necessarily duplicated in performing numerous, widely different processes chemically considered."

The Court of Customs and Patent Appeals (33) has held: ". . . the applicant should claim his invention as a process, which is the only way a 'use' can be claimed."

The reasoning in allowing process claims is set forth in various decisions (22, 36, 42).

The difficulty with process claim protection is twofold: The first is the location in which the process is being carried out, and the second is who is carrying out the process. If an inventor has a patent on a process for making an old compound and that process is practiced in one country and the product itself imported into a different country, different laws apply in different countries. The International Association for the Protection of Industrial Property recommends that the patentee should have the same rights on the imported product which is accorded him by the domestic law in the country of importation had the processes been practiced there. As a practical matter prosecution against a source outside of a country is more complicated and more costly than against a domestic infringer of a process patent. The Tariff Act offers some aid (*52*).

Lyon v. General Motors Corp. (*30*) held:

"Although Defendant, a Delaware corporation, admittedly has a place of business in this district, it has manufactured no wheel covers by the patent method, here, within the period alleged in the complaint. Therefore, no actionable infringement occurred here.

"The sale of the product of the infringing process or method would not constitute an infringement of the method patent within this district. (Merrill v. Yeomans, 94 U.S. 568 (1876); In re Amtorg Trading Corporation, 75 F.2d 826, 24 USPQ 315 (CCPA 1935); Foster D. Snell, Inc. v. Potters et al., 88 F.2d 611, 33 USPQ 112 (C.C.A. 2, 1937); Metro-Goldwyn-Mayer Corporation v. Fear, 104 F.2d 892, 42 USPQ 101 (C.C.A. 9, 1939); 3 Walker on Patents, Deller's Edition, § 843 (1960 Supp.); 69 C.J.S., "Patents", § 289)."

The action for unfair competition and patent infringement was then dismissed for lack of jurisdiction.

Thus, this case holds that in a suit to protect a process patent, the process must have been not only practiced in the United States, but the suit must be brought in the District Court for the district in which the infringement occurred. The legal aspects of jurisdiction, service of parties, joinder of causes of action, and related legal technicalities can raise problems.

The second difficulty with process claims for new uses of an old chemical is: Who should be sued? That is: How can the patent be enforced? If a patent issues on a method of treatment involving the administration of an old substance as a medicine, is the actual infringer the physician or the patient? Is the answer different if the "patient" is an adult, a baby, or an animal? If the use is of an insecticide, the infringer is the farmer. If the patentee must file a separate suit for each patient or farmer, he may have a perfectly valid cause of action; but it is illusory protection as it is difficult to catch each separate infringer and if he does, the recovery from each could be mockingly small.

Some protection can be obtained under Section 271, which provides (*54*) in part:

"(b) Whoever actively induces infringement of a patent shall be liable as an infringer.

"(c) Whoever sells a component of a patented machine, manufacture, combination or composition, or a material or apparatus for use in practicing a patented process, constituting a material part of the invention, knowing the same to be especially made or especially adapted for use in an infringement of such patent, and not a staple article or commodity of commerce suitable for substantial noninfringing use, shall be liable as a contributory infringer."

This section of the law was added in the July 19, 1952, revision of the Patent Law, effective Jan. 1, 1953.

The doctrine of contributory infringement has had its ups and downs and presents, in this connection, some very interesting problems between the extreme situations. The old substance can be an ordinary article of commerce, such as sodium chloride, or worse yet, ethyl alcohol. Either of these substances can be used to infringe several patents, and yet the seller may not know the purpose of the sale. What does it take to show "active inducement of infringement"? With ethyl alcohol there is the additional question of the alcohol tax laws, and, by analogy, how close does a manufacturer have to be to the bootlegger to be a contributor to the crime of bootlegging? The other extreme is a compound which is known only in the sense that it is an entry in *Chemical Abstracts* based on an obscure journal somewhere, perhaps even as a result of a misprint; the compound has no other known use; and any sale would obviously be for purposes of inducing infringement. The lines of demarcation between these two extremes are vague, and the decisional law has not yet been the subject of complete development under the new act.

While the answers are extremely important in considering the effective protection available from a process claim, a comprehensive survey of the trend of decisions would be unduly long. For any particular fact situation, the notes in 35 U.S.C.A. 271 (*51*) give a good start. (The annotations show by short summaries each case decided involving a particular section of the statutes.)

One phase of contributory infringement is the subject of an article by William M. Hogg (*25*). Another is involved in Dr. Salisbury Laboratories v. I. D. Russell Co. (*12*); yet another is Fromberg v. Thornhill (*18*).

Subject Matter of Process Claims

An important change in the earlier law is the overruling of the old Brinkerhoff decision (*5*). In Ex parte Scherer (*43*), the claim involved reads as follows:

"29. The method of injecting fluids into the human body comprising the steps of placing a container of an injecting instrument having a jet orifice tightly against the epidermis to provide a hydraulic seal between the edge of the orifice and the epidermis, displacing liquid from the container through the jet orifice at a pressure sufficiently high to produce a jet velocity which causes the jet to puncture the epidermis and penetrate the body tissues therebeneath, including the steps of continuing the high pressure acting on the jet until it has reached a desired depth, and abruptly stopping the high pressure and thereafter continuing the jet at a lower pressure after the high pressure has been exerted and until the liquid has been dispersed at such desired depth."

The Board of Appeals, in part, held:

"[3] A basic question involved in this case is whether methods in which the subject matter treated is the human body and the object of the method is some medical or surgical purpose are within the field of subject matter capable of being patented. It is our opinion that it cannot be categorically stated that all such methods are unpatentable subject matter merely because they involve some treatment of the human body. Claims involving treatment of the human body have been allowed on appeal, see Ex parte Wappler, 26 USPQ 191, and in Ex parte Kettering, 35 USPQ 342. There is nothing in the patent statute which categorically excludes such methods, nor has any general rule of exclusion been developed by decisions.

"We do not believe that Morton v. The New York Eye and Ear Infirmary is sufficient to establish the principle that all methods involving treatment of the body are thereby not patentable. The patent in that case was held invalid because the material used in the method was old, the step of inhaling the material was old, and the material had been inhaled by persons before; in other words, all aspects of the method, the procedure, the material used, and the subject treated, were old in combination, and the novelty consisted solely in the discovery of the effect produced."

A next step is that of the Board of Appeals in Ex parte Zbornik and Peterson (*61*), in which Claim 25 under consideration reads:

"A process of treating Air Sac Infection in fowl which comprises introducing into the intestinal tract of the bird infected with the causative agent of said disease a poultry feed containing approximately 0.1% of a compound selected from the group consisting of para-aminobenzoic acid, water-soluble salts of para-aminobenzoic acid and mixtures thereof, and maintaining said treatment for a period of not less than five days."

Even though *p*-aminobenzoic acid had been fed to fowl previously, a publication reporting a test series in which the acid was administered as a control test for malaria in ducks, the board gave great weight to the limitation that the fowls were infected with air sac infection and allowed the claims.

Examples of claims which have appeared in patents are as follows: Claim 10 of a patent to Gysin and Knusli (*21*) reads:

"A method of inhibiting the growth of a plant which comprises bringing into contact with at least a part of the plant an agricultural composition consisting essentially of a triazine derivative of the formula

$$\begin{array}{c}\text{Br}\\|\\ \text{triazine ring with N at positions 1,3,5; X and Y at 4,6}\end{array}$$

wherein X is a member selected from the group consisting of ethylamino and isopropylamino groups, and Y is a member selected from the group consisting of methylamino, ethylamino, diethylamino, n-propylamino and allylamino groups, X and Y being different from each other, in a concentration sufficient to inhibit plant growth."

Claim 4 of a patent to Willard and Maitlen (58) reads:

"The method of combatting fungi in the soil which comprises applying a fungicidal amount and concentration of 1-chloro-2-nitropropane to the soil."

Claim 1 of a patent to Hofmann and Troxler (24) reads:

"The method of treating mental disturbances of neurotic and psychic origin, which comprises administering a therapeutically effective dose of a compound, having psychic stimulant properties, of the formula

$$\text{indole with O-CO-R substituent and } CH_2-CH_2-N(\text{lower alkyl})_2 \text{ side chain}$$

wherein R is a member selected from the group consisting of lower alkyl and phenyl groups."

Claim 7 of a patent to Didusch (10) reads:

"A method of closing an incision in a kidney which consists in wrapping a ribbon of absorbable material around said kidney, approximating the edges of the incision, securing the ends of the ribbon to maintain the edges of the incision in approximation, and embedding the kidney and ribbon in living tissue to permit complete absorption of the ribbon and healing of the incision."

Claim 1 of a patent to Steinberg (47) reads:

"In the treatment of verrucae, the step which comprises injecting directly into the verrucae a vitamin A compound of the group consisting of vitamin A and the fatty acid esters thereof."

Claims 3 and 5 of a patent to Burggraf-Brockelmann and Strandskov (6) read:

"3. A method of inhibiting the micro-biological growth of *Lactobacillus pastorianus, Pediococcus damnosus* and secondary yeast in beer, which comprises incorporating the antibiotics of polymyxin, terramycin and thiolutin in amounts of from about 3.0 to about 5.0 gamma per milliliter of finished beer."

"5. Beer containing thiolutin in an amount of about 3.0 gamma per milliliter and polymyxin in an amount from about 1.0 gamma to about 3.0 gamma per milliliter, whereby micro-biological growth of *Lactobacillus pastorianus, Pediococcus damnosus* and secondary yeast is inhibited."

The District Court held in Bancroft v. Watson (2) in allowing process claims:

"[3] The question arises, however, whether on this point the law has been changed by the 1952 codification of the patent laws. This statute is, in effect, more than merely a codification, but introduces some new provisions into the law of patents. 35 U.S.C. § 100, subsection (b), in defining the word 'process', for the first time, provides that the term
"'* * * includes a new use of a known process, machine, manufacture, composition of matter, or material.'
"In other words, a new use of a hitherto known process or composition of matter may be patentable, provided, of course, all the other requisites of patentability are met. The mere fact that what the inventor seeks to patent is a new use of a previously known invention is no longer a bar to a patent . . .
"It is sufficient to apply Section 100(b), above quoted, which, in effect, has abrogated the principle laid down in a series of cases to the effect that a new use of an old process or an old device is not patentable on that ground alone. What we are confronted with in the two process claims is not a new process, as such, but the application of a previously known chemical to an entirely new and different use, and making it a participant in the process. There do not seem to have been very many decisions construing the new provision to which reference has been made, but there are two cases decided by the Court of Customs and Patent Appeals in regard to the matter that appear to be pertinent."

There is a risk that the claims will not recite all of the necessary steps and that the claims thereby fail. Ex parte Salathiel (41) held:

"[1] The doctrine that the nature of the chemicals used in a process should not be ignored in considering patentability of such process appears well established. However, the proposition that merely mixing chemicals or ingredients to produce a composition is, in general, unpatentable as an obvious way of effecting the composition, has not been overruled [2] In the present case the claims do not specifically call for contacting the mud mixture with the well wall. Accordingly we do not consider that the actual use of the composition is involved. Therefore the claims, unlike those allowed in Ex parte Wagner, merely involve the mixing of certain ingredients. Since the claims deliberately omit the step which the Board in Ex parte Wagner held to distinguish over the Wayne case, we arrive at the conclusion that the allowance of the present case is not warranted under the Ex parte Wagner decision; . . ."

The claims on appeal were in Jepson (27) form reading:

"16. In a process for drilling a well into subsurface formations with rotary drilling tools wherein there is circulated in the well a water-base drilling mud containing colloidal particles of clayey material suspended in sufficient water to render the same circulatable, the method of transforming said water-base drilling mud into an oil-in-water emulsion drilling fluid having little tendency to lose water contained therein into surrounding earthen formations which comprises admixing with said drilling mud a hydrocarbon oil, . . ."

Hence, care must be used to recite clearly as positive limits those steps which distinguish from the prior art.

Manufacture

Some new uses of old chemicals can be protected by claiming as an article of manufacture, based on a particular disposition or a special form as adapted for particular uses (56).

A patent to Greif (19) has somewhat more structure. Claim 1 reads:

"An oral pharmaceutical preparation having a prolonged release comprising a plurality of medicament granules, substantially all being from 12 mesh to 80 mesh, each coated with a layer of water insoluble, partly digestible hydrophobic material, the thickness of coating varying directly with particle size whereby in oral use the very fine granules rapidly release their medicament and the granules of increasing size release their medicament more and more slowly."

A patent to Consolazio (9) has still more structure. The claim reads:

"An internally reinforced sodium chloride tablet comprising compressed granules of sodium chloride; and an internally disposed cellular stroma of a thin, permeable, dialyzing film of a material selected from the group consisting of cellulose acetate and cellulose nitrate, the cells of said stroma containing said granules of sodium chloride whereby the sodium chloride is rendered slowly available when the tablet reaches the gastro-intestinal tract, the solution time of the sodium chloride in said tablet in the gastro-intestinal fluids being from 60 to 80 minutes for a ten grain tablet."

There are many other patents on structures involving new uses in particular forms. These examples are illustrative, but not definitive.

Current Trends

In Phillips Petroleum Co. v. Ladd (*35*), the District Court held Claims 17 and 19 allowable:

"17. A rubbery polymer of 1,3-butadiene characterized by at least 85 per cent cis-1,4-addition."

The prior art mentioned by name a 100% cis polymer and said the polymer could not be made. Polymers are not defined clearly by Geneva nomenclature. Hence, even though a polymer arguably within the class was named, the claims were allowed. The court held:

"A mere naked formula for a chemical compound which teaches the art nothing about the product which it may represent, and does not put anyone in possession of the invention, is not the type of statement that should be relied upon for anticipation."

The court distinguishes Von Bramer (*55*) and cites several cases.

Phillips Petroleum Co. v. Ladd should not be classed as a new use of an old compound, but rather as a patent on a new polymer.

Patent problems on what are new polymers of a conventional monomer must follow the chemical problems of characterization and description of these polymers.

Several different approaches to claiming an invention were used in the prosecution of In re Riden and Flavin (*38*). The Court of Customs and Patent Appeals sustained the rejection of claims drawn to compounds and also of claims drawn to novel compositions containing these compounds.

The court reversed the Patent Office and allowed process claims. Claim 21 reads:

"A process for protecting a material from attack by a member of the group consisting of microorganisms and nematodes comprising applying to said material an effective amount of a halogenated ethenyl sulfone of

$$R-\underset{\underset{O}{\|}}{\overset{\overset{O}{\|}}{S}}-\underset{X}{\overset{}{C}}=\underset{Y}{\overset{}{C}}-Z$$

wherein R is an alkyl group, at least two of X, Y and Z are halogen of atomic weight not over 80 and the third member of X, Y and Z is selected from the group consisting of halogen of atomic weight not over 80 and hydrogen."

In part the Court held:

"[6] We are of the opinion that appellants have invented a new use for a known compound (or at least a group of compounds including known compounds). In claim 21 this new use has been properly claimed as a process under 35 U.S.C. 100(b). Appellants were the first to use the compounds of Boehme et al. as pesticides. We do not feel that such a use would be obvious, to one having ordinary skill in the art, from the teaching of Metivier because of the differences between the respective compounds. The mere fact that the aliphatic sulfones of appellants have fewer carbons and hydrogens than the sulfones of Metivier which have a ring structure does not suggest that these compounds would have similar utility. There is a considerable degree of unpredictability in the pesticide art."

It would appear that the useful properties of the novel compounds did not distinguish from the unknown but similarly useful properties of old compounds.

This case is an excellent review of situations involving new uses of a series of related compounds, some new and some old, where claims are presented to the new compounds, and all are grouped in the method claims.

Analysis of the decision emphasizes the importance of process claims in any application dealing with novel uses of compounds, some of which may turn out to be previously known.

This decision quotes from In re Mills (*31*) to bring out the key point:

"We do not stop to consider the correctness of these two decisions, but merely caution against the tendency 'to freeze into rules of general application what, at best, are statements applicable to particular fact situations'."

Patents on Uses—Historical

With changes in the law, the propriety of certain forms of patent claims changes from time to time. Before 1900 it was common to have a claim on a use as such. Patents on uses as such had been going out of favor for some time. The final blow was In re Thuau (*50*). A claim therein reads:

"A therapeutic product for the treatment of diseased tissue, comprising a condensation product of metacresolsulfonic acid condensed through an aldehyde."

The particular compound itself was old. The Court of Customs and Patent Appeals reached the conclusion that patents should not be granted for old compositions. A comprehensive discussion involving In re Thuau and many other cases appear in an article by Ryan (*39*). This article cites among others, an article by Tashof (*49*). The conclusions reached by Mr. Tashof are out of date to the extent that now patents are granted on methods of treating the human body—Ex parte Scherer (*43*).

The Ryan article mentions six cases in which use patents were upheld by the U. S. Supreme Court. All were prior to 1902. A product claim phrased as a use is obsolete.

In summary, patents, creations of statutes, are subject to human frailties in preparation, prosecution, and interpretation. Some cases cannot be reconciled and some, which while once good law, are out of date. For the chemist who has developed a new use for an old compound, the first step should be to ascertain if there is some change in form, purity, carrier, or other feature which can distinguish the product from that which was old. In other words, if the two products were put in two bottles, side by side, would there be a difference, and is that difference important? If not, protection may be obtainable by process claims under 35 USC 100 *(51)* and direct or contributory infringement under 35 USC 271 *(54)*.

Additional protection should be considered as an article of manufacture if the invention can be phrased in terms of such an article.

The real question is whether the new concept would have been obvious to one skilled in the art, 35 USC 103 *(53)*. If it is clearly patentable, this article may help in claiming the invention adequately. If the concept would have been obvious, it is not patentable, and no amount of rephrasing nor change in claim form can cure the basic defect.

Literature Cited

(1) Ajax Metal Co. v. Brady Brass Co., 155 Fed. 409; (D.Ct.N.J. 1907); 160 Fed. 84 (3rd Cir. 1908).
(2) Bancroft v. Watson, 170 F.Supp. 78, 120 U.S.P.Q. 265 (D.Ct. D.C. 1959).
(3) Biesterfeld, Charles H., "Patent Law," 2nd ed., pp. 59-62, John Wiley & Sons, New York, N.Y., 1949.
(4) Biesterfeld, Charles H., supra, pp. 63-78.
(5) Brinkerhoff, 24 Manuscript Decisions 349. (Available in U.S. Patent Office files only.)
(6) Burggraf-Bockelmann, John, and Strandskov, F. B., U.S. Patent **2,798,811** (July 9, 1957).
(7) Chamberlain, G. D., Ex parte, 1931 C.D. 10; 413 O.G. 1101, at 1102; 2 U.S.P.Q. 145 (P.O. Bd. App. 1929).
(8) Cochrane v. Deener, 94 U.S. 780, at 788 (U.S. Sup. Ct. 1876).
(9) Consolazio, W. V., U.S. Patent **2,478,182** (Aug. 9, 1949).
(10) Didusch, W. P., U.S. Patent **2,143,910** (Jan. 17, 1939).
(11) Dr. Salisbury's Laboratories v. I. D. Russell Co., 97 F. Supp. 695; 90 U.S.P.Q. 247 (D.Ct. W.Mo. 1951).
(12) Dr. Salisbury's Laboratories v. I. D. Russell Co., 212 F.24 414; 101 U.S.P.Q. 137 (8th Cir. 1954).
(13) Dr. Salisbury's Laboratories v. I. D. Russell Co., 121 F. Supp. 709; 101 U.S.P.Q. 212 (D.Ct. W. Miss. 1953).
(14) Farbenfabriken of Elberfeld Co. v. Kuehmsted, 171 Fed. 887; (C.Ct. Ill. 1909).
(15) Ferguson, E. A., Jr., U.S. Patent **2,486,937** (Nov. 1, 1949).
(16) Fisher, In re J.D., 50 C.C.P.A. 1023; 307 F.2d 948; 1962 C.D. 639; 785 O.G. 379; 135 U.S.P.Q. 22, 26 (C.C.P.A. September 1962).
(17) Fisher, In re J.D., 50 C.C.P.A. 1019; 314 F.2d 817; 789 O.G. 1145; 137 U.S.P.Q. 150 (C.C.P.A. March 1963).
(18) Fromberg, Inc. v. Thornhill, 315 F.2d 407; 137 U.S.P.Q. 84 (5th Cir. March 1963).
(19) Greif, Martin, U.S. Patent **3,078,216** (Feb. 19, 1963).
(20) Gruskin, Benjamin, U.S. Patent **2,120,667** (June 14, 1938).
(21) Gysin, Hans, and Knüsli, Enrico, U.S. Patent **3,078,154** (Feb. 19, 1963).
(22) Hessel, Ex parte, 137 U.S.P.Q. 384 (P.O. Bd. App. March 1962).
(23) Hoffman, Felix, U.S. Patent **644,077** (Feb. 27, 1900) cited in Farbenfabriken of Elberfeld Co. v. Kuehmsted, supra *(14)*.

(24) Hofmann, Albert, and Troxler, Franz, U.S. Patent 3,078,214 (Feb. 19, 1963).
(25) Hogg, W. N., J.P.O.S. 42, 683-93 (October 1960).
(26) Hoskins v. General Electric, 212 Fed. 422 (D.C. Ill. 1914) 224 Fed. 464 (7th Cir. 1915).
(27) Jepson, Ex parte, 1917 C.D. 62; 243 O.G. 525 C.D. 1917.
(28) Kuehmsted v. Farbenfabriken of Elberfeld Co., 179 Fed. 701 (7th Cir. 1910).
(29) Laurence, Dean, ADVAN. CHEM. SER. No. 46, p. 73 (1964).
(30) Lyon v. General Motors Corp., 200 F.Supp. 89; 131 U.S.P.Q. 310 (D.Ct. Ill. 1961).
(31) Mills, In re Victor, 47 C.C.P.A. 1187, 1190; 281 F.2d 218; 761 O.G. 286; 1960 C.D. 523; 126 U.S.P.Q. 513, 517 (C.C.P.A. 1960).
(32) Morehouse, N. F. and Mayfield, O. J., U.S. Patent 2,450,866 (Oct. 5, 1948).
(33) Papesch, In re Viktor, 50 C.C.P.A. 1084; 315 F.2d 381; 794 O.G. 14; 137 U.S.P.Q. 43 (C.C.P.A. March 1963).
(34) Pfeiffer, F. L., U.S. Patent 3,501,721 (Aug. 28, 1862).
(35) Phillips Petroleum Co. v. Ladd, 219 F.Supp. 366; 138 U.S.P.Q. 421 (D.Ct. D.C. July 1963).
(36) Pieroh and Werres, In re, 50 C.C.P.A. 1471; 319 F.2d 248; 797 O.G. 6; 138 U.S.P.Q. 239 (C.C.P.A. June 1963).
(37) Radio Position Finding Corp. v. Bendix Corp., 205 F.Supp. 850; 133 U.S.P.Q. 638 (D.Ct. Md. 1962).
(38) Riden and Flavin, In re, 50 C.C.P.A. 1411; 318 F.2d 761; 138 U.S.P.Q. 112 (C.C.P.A. June 1963).
(39) Ryan, Martin A., J.P.O.S. 29, 787-821 (1947).
(40) Rystan Co. v. Warren-Teed Products Co., Inc., D.Ct. N. Tex. 1952; 105 F.Supp. 56; 92 U.S.P.Q. 419.
(41) Salathiel, Ex parte, 106 U.S.P.Q. 419-420 (P.O. Bd. App. Oct. 30, 1953).
(42) Santmyer, Ex parte, 132 U.S.P.Q. 202 (P.O. Bd. App. February 1960).
(43) Sherer, Ex parte, 103 U.S.P.Q. 107 (P.O. Bd. App. July 23, 1954).
(44) Shell Development v. Watson, 148 F.Supp. 373; 112 U.S.P.Q. 313 (Feb. 14, 1957); D.Ct.D.C.
(45) *Shepards Federal Reporter Citations*, Shepard's Citations, Inc., Colorado Springs, Col., **1**, 1938; **2**, 1938-1953; Supp. 1953-1961 and monthly.
(46) Steelmand, Ex parte S. L., and Kelly, T. L., 798 O.G. 259 (P.O. Bd. App. March 1962).
(47) Steinberg, M. D., U.S. Patent 2,760,901 (Aug. 28, 1956).
(48) Tanczyn, In re, 40 C.C.P.A. 886; 202 F.2d 785; 672 O.G. 7; 97 U.S.P.Q. 150 (C.C.P.A. March 1953).
(49) Tashof, I. P., *Chem. Eng. News* **25**, 491-94 (1947).
(50) Thuau, In re (1943), 30 C.C.P.A. 979; 135 F.2d 344; 554 O.G. 14; 57 U.S.P.Q. 324 (C.C.P.A. 1943).
(51) U.S.C.A., West Publishing Co., St. Paul, Minn., particularly Title 35, Patents (1954) with yearly cumulative annual pocket supplements.
(52) U.S.C. **19**, Sec. 1337a.
(53) U.S.C. **35**, Sec. 103.
(54) U.S.C. **35**, Sec. 271.
(55) Von Bramer and Ruggles, A.C., In re, 29 C.C.P.A. 1018; 127 F.2d 1, 49; 542 O.G. 183; 53 U.S.P.Q. 345 (C.C.P.A. 1942).
(56) Widmann, R. R., U.S. Patent 2,628,184 (Feb. 10, 1953).
(57) Widmann, R. R., supra, Statutory Notices (35 U.S.C. 290) to patent suits involving Patent 2,628,184, 671 O.G. 906; 675 O.G. 553; 694 O.G. 560; 670 O.G. 914; 673 O.G. 866; and 691 O.G. 10.
(58) Willard, J. R., and Maitlen, E. G., U.S. Patent 3,078,209 (Feb. 19, 1963).
(59) Wiswall, R. H., Jr., U.S. Patent 2,486,351 (Oct. 25 1949).
(60) Yale and Bernstein, Ex parte, 119 U.S.P.Q. 256 (P.O. Bd. App. April 1958).
(61) Zbornik and Peterson, Ex parte, 109 U.S.P.Q. 508 (P.O. Bd. App. March 20, 1956).

---

[1] Legal abbreviations are defined on page viii.

RECEIVED October 10, 1963.

10

Patentability of Natural Products, Plant Isolates, and Microbiological Products

JOHN H. SCHNEIDER
Johnson & Johnson, New Brunswick, N. J.

> There are operations of the law of nature which are not patentable. On the other hand, there have been discoveries made in scientific fields which in a sense are phenomena of old and well-known products. These are not described as laws of nature. Problems arise chiefly in relationship to what already is known. Failure to distinguish between product coverage in this area resulted in a categorical application of the rule to both situations. This has occasioned much of the misunderstanding. The courts never intended the meaning of the rule which would declare otherwise patentable compositions of matter to be unpatentable merely because their creation might be the handiwork of nature rather than of man.

The advent in recent years of important new commercial products having their origin in plants or as a result of microbiological fermentation has given rise to further consideration of the meaning of the terms "principle of nature" and "products of nature." In medicine, in particular, the era of antibiotic developments, starting with the penicillins and progressing through many highly useful antibiotics, such as erythromycin and tetracycline, has been one of the most productive periods in the history of medical progress. Its benefits to mankind are without question of great importance. Had its products and developments been construed as unpatentable, it is doubtful that investment capital would have been so readily available for the extensive research and development programs necessary for their achievement.

In origin, the prohibition of the patentability of principles of nature constituted an effort on the part of the courts to compel the inventor to define his invention in real and tangible form rather than as an intangible principle or mode of operation. The prohibition of products of nature appears to stem from the elementary proposition that one may not patent what already exists and is known.

When a chemical structure is formed, it is possible to conceive that a principle of nature is involved in the act of the atoms combining to form a particular structure. In other words, when a given environment is provided, the atoms will go together in one and only in one way to form a given structure. By making changes in environment, it is possible to alter circumstances to give rise to a different combination of the same atoms to form a different structure. In the papers by Ruby (9), starting in 1940 and extending through 1943, the theory propounded would render chemical product claims invalid. The basis for this is the concept that a principle of nature is involved in the structure itself. This is generally conceded today to be erroneous. While a principle of nature may be regarded as being involved in the act of combining atoms to form a particular novel structure, once the structure is thus deliberately formed by man, the structure itself is not a principle of nature, nor is it a product of nature.

Although a principle of nature in itself is not patentable, the utilization of a principle of nature to accomplish a given purpose is and has always been patentable. Hence, one who discovers a new process for making a compound, whether new or old, provides a set of steps or conditions which, if pursued, could be said to give rise to the operation of a principle of nature, whereby atoms will combine in a particular way to form a particular structure. This is a proper utilization of a principle of nature, and the process involving the steps or conditions should be patentable. If the structure thus formed is a new composition of matter, it also should be patentable as such.

Principle of Nature Cases

In the Morse Case (7), one of the claims had been worded so that it claimed the effect—the principle of nature by which the inventor's apparatus operated—rather than the process or machinery necessary to produce the desired effect. The court ruling in this instance can be said to be no more than a ruling that the abstract principle discovered is not of itself the patentable part of invention. It is the concrete embodiment or the manner of application, either as a process or in a machine—the "how to use" to the benefit of mankind—which is patentable.

The net conclusion to be drawn from the decisions on unpatentability of principles of nature is that the courts were merely endeavoring to exclude from patentability the intangibles of the invention—the base principles of operation itself. There is no evidence of an intent to establish a classification of patentable subject matter which would preclude claims to novel substances of natural origin on the sole ground that nature may have played a greater part in their creation than did the inventor.

Product of Nature Cases

In the General Electric tungsten case (*4*), the court held that a broad product claim to pure ductile tungsten was invalid as a claim to a product of nature. The evidence in the case indicated that while pure tungsten had been known, it had never been known to exist in a ductile form. The process discovered by the inventor consisted of a heating and drawing procedure which brought out a natural property of ductility inherent in the metal under certain circumstances. In the decision, the court made the following statement:

"If it (tungsten) is a natural thing, then clearly, even if Coolidge was the first to uncover it and bring it into view, he cannot have a patent for it because a patent cannot be awarded for a discovery or for a product of nature or for a chemical element."

This statement was not the basis for the holding of invalidity but has been cited often. Actually there is little if any authority for the statement.

Chemical Compound Cases

There are numerous decisions that hold product claims to chemical compounds as "compositions of matter" to be valid within the meaning of the patent statutes. With few exceptions there are no specific references to the rule that products of nature are unpatentable. The statements in the decisions are in sharp contrast with the thought that the rule was at all applicable. For example in the Kuehmsted case (*6*), the pure chemical compound known generally as "aspirin" was held to be patentable because of its new utility. The court held that whatever may have been its antecedents chemically, aspirin in pure form was a new thing. The formerly known crude material was legally a different material in that it had no utility as a medicinal. The court held that aspirin was an article of manufacture within the meaning of the patent law.

Aside from the statutory requirements of utility, the other important test of patentability is novelty or unobviousness—whether the substance actually existed in nature and whether or not its isolation was obvious from the prior art. Reverting to the tungsten and aspirin cases, tungsten did exist as such in nature whereas aspirin existed previously only as part of a medically unusable crude composite.

The question is frequently asked: "Is the mere isolation of a hitherto existing but hidden substance sufficient basis for patentability, or is it necessary that it never have existed?" While it has been held that mere purification of a known substance will not support a product claim to the improved product, the courts have definitely held that isolation of a hitherto unknown substance may be sufficient if the end result as a whole is not obvious from the prior art. Adrenalin when extracted from gland tissue was held to be patentable. The court (*8*) in holding the patent valid made the following statement:

"But even if it (adrenalin) were merely an extracted product without change, there is no rule that such products are not patentable. Takamine was the first to make it available for any use by removing it from the other gland tissue in which it was found, and, while it was of course possible logically to call this a purification of the principle *(sic)*, it became for every practical purpose a new thing commercially and therapeutically. That was good ground for a patent."

In like manner, in the Kuehmsted case (*6*), which concerned aspirin, the court made the following statement:

"Hoffman has produced a medicine indisputably beneficial to mankind—something new in a useful art, such as our patent policy was intended to promote. Kraut and his contemporaries, on the other hand, had produced only, at best, a chemical compound in an impure state. *And it makes no difference, so far as patentability is concerned, that the medicine thus produced is lifted out of a mass that contained, chemically, the compound;* for, though the difference between Hoffman and Kraut be one of purification only—strictly marking the line, however, where one is therapeutically available and the others were therapeutically unavailable—patentability would follow. In one case, the mass is made to yield something to the useful arts; in the other what is yielded is chiefly interesting as a fact in chemical learning."

Plant Isolate Cases

In contrast to the aspirin and adrenalin cases, there is a line of cases holding that the mere extract of a substance from a natural environment to produce a pure or more stable substance does not give rise to a valid product claim. For example, in the case of In re King (*5*), the court held unpatentable the claims to a therapeutic product that were couched in terms amounting to a definition of pure vitamin C. The evidence showed that this material had been present for years as a component of lemon juice, although it was not known by the name of vitamin C or hexuronic acid C. The appellants in that case contended that it was invention to have discovered that hexuronic acid C is vitamin C. The court stated that, had the substance not been known before it was isolated by the appellants, there would be force to that contention. All the appellants did, however, was to produce a compound that was old in the art although not recognized by the same name. Its properties were the same as those of the material already known in nature.

The case of Ex parte Berkman and Berkman (2) which was decided by the Patent Office Board of Appeals concerned claims to a carotene product isolated from raw plant source without substantial change except as to purity and stability. The claims were held to be unpatentable to the applicants. The applicants had admitted in the record of the case that the substance thus extracted was identical with that which occurred in the raw material. It was noted by the board that the same substance was stable in its natural environment as long as the raw material remained alive but became unstable soon after the plant source of the raw material died. It was apparently for this reason that the argument of stability in the isolated material failed to impress the board that there was any difference between the isolated material and the material as it existed in the plant source. In this case the board distinguished over the adrenalin case by contending that the adrenalin present in the suprarenal glands was extremely dangerous for injection into the body whereas the isolated material did not give rise to this problem. Hence, for the first time the material known as adrenalin was made available in an injectable form.

The case of Ex parte Snell (10) involved a physiologically active compound identified thereafter by the structural formula of pyridoxamine. The evidence in this case showed that one form of vitamin B_6 present in yeast and extracted as the predominant constituent was pyridoxamine, although other substances were present. The compound had been synthesized by the appellant, and the synthetic compound was identical in structure with the compound found in nature. It is not surprising therefore that the board rejected the claim as being directed to a naturally occurring product—that is, to a product that was not "new" in the contemplation of the law. In many of these cases, even though the product was not deemed patentable, the method of isolating the product was found to be patentable. For the product under these circumstances to be patentable, there must be a modification of its properties such that the product will produce a result which was not possible with the unisolated product.

A rather interesting case on the subject of patentability of plant isolates (3) involved an extract made from the root of the cube plant. It was effective in paralyzing insects and other forms of life which may be exterminated by contact poison. The court in this decision made the following statement:

"We believe it to be a sound pronouncement to say . . . The discovery of a natural phenomenon, or of a quality or attribute of a well-known article, which discovery is of value to mankind, may be entitled to patent protection. The objection frequently offered to the patentability of such a discovery is that it is a law of nature or a principle of nature and for that reason not patentable. Section 31, Title 35, U.S.C.A. (now 35 U.S.C. 102) authorizes the issuance of a patent to

"'Any person who has invented or discovered any new and useful art, machine, manufacture, or composition of matter, or any new and useful improvements thereof . . . not known or used by others in this country, before his invention or discovery thereof, and not patented or described in any printed publication in this or any foreign country, before his invention or discovery thereof . . .'

"There would seem to be no valid reason or sound support for a position which would deny to discoveries by researchers in the field of science the protection of our patent laws when such discovery is that an old, or at least well-known chemical product, will, acting in a given state, alone, or combined with other elements or physical elements, produce new, unknown, and unexpected results, whereas one who puts together at least two old and well-known chemical substances in certain prescribed proportions and gets new results helpful to man may receive patent protection. In the latter case, patent protection is universally accorded to the discovery."

The appellate court in this decision, in reversing the lower court, denied the patentability of the cube root extract as a product on the ground that the properties of the extracted material were the same as the properties of the same material in its natural environment in the unextracted cube root. There is nothing inconsistent in this position. At the same time the court made the following statement:

"A discovery in the field of science of a new quality or phenomenon of an old product may be (other necessary facts such as being first, timely application, etc., existing) the proper subject of a patent. It does not fall within the term 'law of nature' as that expression is used in the patent."

Of course, the discovery of a new quality of an old product does not make the product itself patentable, but invention may be claimed in a new process of using the product—for example, in applying it to vegetation as an insecticide or in a new composition of matter involving its admixture with a carrier which makes it suitable for the novel purpose (*11*).

Microbiological Product Cases

In the so-called "tetracycline case" (*1*), there was evidence that in the preparation of aureomycin broth by the fermentation of S. *aureofaciens*, some small percentage of tetracycline was coproduced. Aureomycin was known and produced by fermentation substantially before the discovery of tetracycline. The argument was raised by the patent examiner when the tetracycline application was in the Patent Office that tetracycline must be produced inherently in the fermentation in the production of aureomycin. The applicant was able to show, however, that the amount of tetracycline produced in the fermentation broth in aureomycin production was so small that it was of inconsequential value to mankind as an antibiotic. In fact, most methods of analysis did not even detect its presence in the broth. On the basis of this showing, claims to tetracycline were allowed.

The key question involved in the situation of this type appears to be one of utility. Even though the prior product has been known and disclosed, if it existed in a form which possessed little or no utility, then a claimant who has reduced it to a purer form which possesses substantial or great utility can secure a patent on it even though it is literally a purer form of an old product. This is the holding of the aspirin, adrenalin, and vitamin B_{12} cases. In all of these, the product had been known to exist in the prior art but possessed little or no utility, whereas the claimed product, while technically a purer form of the old product, possessed great and new utility and hence was held to be different in kind rather than in degree or, in terms of the present law, to be unobvious from the prior art. In contrast to this, numerous decisions hold that a purer form of an old product differing only in degree, in other words, possessing better utility of the same kind, is not patentable because it has been previously disclosed, and the new product is merely an obvious variation of the old as far as utility is concerned.

Examine for a moment the situation which exists when an organism placed in a nutrient medium and allowed to grow produces, as a metabolic by-product, a substance found to possess so-called "antibiotic" properties of a useful nature. One can assume that the same organism in its natural environment in the soil must produce the same metabolic by-product. However, it is difficult to conceive of the utility of a soil sample containing this by-product as having any useful medicinal value. Thus while the antibiotic substance may be present in the soil sample, it might just as well not be in existence insofar as its availability to mankind for arresting infections is concerned. Hence, there is no real problem in finding product patentability in a situation of this type.

Conclusion

In spite of some unfortunate statements, as in the tungsten case, the law never intended a rule which would declare otherwise patentable compositions of matter to be unpatentable merely because their creation might be conceived to be the handiwork of nature rather than of man. The courts merely desired in the principle-of-nature cases to force the inventor to claim only the tangible manifestations of his contribution. As to those cases where products have been declared unpatentable, the courts may be said to have been looking primarily to the test as to the presence or absence of novelty or invention.

Modern trends in the chemical and medical sciences with the wide development of synthetic substances, the development of highly important antibiotics, and the discovery of new hormones furnish strong arguments in favor of patentability of such products. Both the Constitution and the patent statutes recognize the patent monopoly grant as beneficial for the promotion of the arts and sciences.

One rather interesting development in recent times is the emphasis placed on the utility and on the unobviousness of the invention. The Patent Act of 1952 (which is recognized officially as being a codification of the case law) particularly stresses the unobviousness factor. In more recent decisions where the utility factor and the unobvious factor have been prominent, the trend is to recognize utility and unobviousness as being of equal significance with structure, so that a patentable discovery is essentially a composite of the novelty of the substance, the nature of the utility, and the unobviousness of the composite to those skilled in the art to which the discovery pertains. A utility of a different kind may be found to be present where a substance, although existing in nature, is in a form or an environment in which the utility is simply not available even though it may be inherent in the substance. The present practice of granting product claims to chemical compounds, to plant isolates, and to microbiological products is not only in accord with established law and the intent of the drafters of the statutes but also in keeping with sound public policy.

Literature Cited[1]

(1) Am. Cyanamid Co., in matter of, Federal Trade Commission, initial decision, Oct. 30, 1961.
(2) Berkman and Berkman, Ex parte, 90 U.S.P.Q. 398 (P.O. Bd. App. 1950).
(3) Dennis v. Pitner 106 F.2d 142 (7th Cir. 1939).
(4) General Electric v. De Forest, 28 Fed.2d 641; (3rd Cir. 1928).
(5) King, In re, 107 F.2d 618 (C.C.P.A. 1939).
(6) Kuehmsted v. Farbenfabriken, 179 Fed. 104. (7th Cir. 1910).
(7) O'Reilly v. Morse, 15 How. 62, 1853 (Sup. Ct.).
(8) Parke-Davis v. Mulford, 196 Fed. 496 (2nd Cir. 1912).
(9) Ruby, Charles E., *Temple Univ. Law Rev.* XV, 27-64 (1940); XV, 321-60 (1941); XVII, 1-63, (1942); XVII, 385-436 (1943).
(10) Snell, Ex parte, 86 U.S.P.Q. 496 (P.O. Bd. App. 1951).
(11) Thuau, In re, 135 F.2d 344 (C.C.P.A. 1943).

---

[1] Legal abbreviations are defined on page viii.

RECEIVED October 10, 1963.

11

Foreign Patent Coverage on Chemicals and Medicinals

ALAN SWABEY

Alan Swabey & Co., Montreal 2, Canada

> Foreign patenting of chemicals and medicinals is governed by rules which differ widely from those of the United States. In most cases, less protection is afforded, but there are exceptions where broader coverage may be obtained. Most foreign countries do not grant patents on chemical products, so these inventions must be protected in process terms or by other methods that differ from those employed in the United States. These and other problems of foreign patents can be dealt with effectively only by a close look at what type of protection each country gives and by trying to fit the invention into the protection.

Because of the highly developed patent law and the large number of patents applied for in the United States, rulings by the Patent Office or the courts are available to establish the patentability of almost any type of chemical invention. Abroad, this is not always so. In the 200 or so foreign territories where patents may be obtained, the laws are generally less sophisticated than they are in the United States and vary widely from one another. So, the exact patent position of a given invention is often uncertain.

A patent attorney in one European country in reply to a questionnaire asking what could be patented there said that because his country lacks an examination procedure, officials are bound to grant almost any patent applied for. He pointed out that only the courts are entitled to decide if a patent is valid or not, if and when the case is submitted to them.

So, it would be misleading to try to lay down any universal hard and fast rules as to where one can patent what. It is, however, useful, if superficial, to identify the main types of patentability requirement and restriction encountered in chemical cases. But this is no substitute for an up-to-date opinion on a given invention from the country where patent protection is desired. Foreign patents cannot be handled by "remote control," nor on the basis of general principles.

Process Patents

The United States Congress has seen fit to grant patents on new chemical substances, regardless of how they are made. A few major industrial countries follow suit, some with limitations as to the field of use. Examples are Great Britain, France, Canada (except foods and medicines), Australia, South Africa, India, and Pakistan. But most countries limit patent protection to processes. Germany and most other European countries are examples. Thus, it may be advisable to patent as many processes as possible leading to a desired end product.

Process patenting abroad, however, may offer greater possibilities than it does in the United States, since the standards of patentability for a process abroad are usually lower than at home. With no product protection, there must be more leniency in process patents, or no one would ever be able to obtain a chemical patent. The Germans, for example, talk about "analogische Verfahren" (or analogous processes). An analogous process is one which may involve ordinary chemical steps, but which, applied for the first time to a particular starting material, results in a product which has unexpected properties. This makes it patentable in Germany, Holland, and many other patentwise nations. In Canada the Supreme Court held, in 1960, in "The Ciba Case" that such a process was patentable. So, the process patent is the primary instrument used abroad to protect many of the inventions on which a product patent would be sought in the United States.

Again, a process patent may give wider protection abroad than it does in the United States. Here, a process patent is not infringed by importation of the processed product. In most foreign countries, the reverse is true. But, here again, there are exceptions. A Spanish patent attorney told the author that, in most cases in which the infringement consists of importing products manufactured under patented processes into Spain, the courts usually reject the claim lodged by the owner of the infringed patent. So, in Spain, a process which may be patentable may in some cases be unprotectable. This is an example of the local ground rules one must look for to get a clear picture of what really can be protected.

Sometimes, an obvious method can be patented because the resulting product has a new use. In Sweden, a patent on DDT was first refused because the applicant claimed "a method of producing certain novel esters of 2,4-dichlorophenoxyacetic acids." The Board of Appeals of the

Swedish Patent Office authorized a patent, provided the claim was changed to read "a method of producing growth-controlling agents." Merely labeling an old process with a new use, as was done in this case, would not qualify for a patent in the United States.

Article Patents

Patents on substances one would normally classify as chemical have been granted abroad because the patentee was alert enough to argue that they were nonchemical. In Sweden, the Patent Office held that a polymer is not a pure chemical compound and so could be patented as and article. In Holland, the Supreme Court ruled that an artificial silk thread has a definite shape and so could be patented as an article. This brought up the problem that patents on processes for producing artificial silk threads (of the type which people had become used to taking out in this field) would no longer cover the article. So, the Swedish Patent Office later rule that, in a case of doubt as to whether a product is a substance or an article, a subclaim can be included in the patent on the process to cover a shaped article made by the process, say paper sheets, fibers, and shaped plastic or metal articles.

The Germans construe the term "chemically produced substance" narrowly. Thus, alloys, mixtures, solutions, mixed crystals, fluorescents, semiconductors, and specific pigments—all mechanically composed—are examples of products on which patents are given even though certain chemical reactions take place in addition to the mechanical mixing process. In Germany again, an insecticide may be patented in a claim which reads "an insecticide, characterized by containing a substance, X, as the effective substance." In Canada, the Patent Office does not consider extraction, where there is no chemical change in the extracted material, to be a chemical process. Nor do the Canadians consider as chemical, a process in which living organisms rather than chemical reagents are used to bring about a chemical change.

Some countries do not allow a patent on a mixture, because of the general rule that a patent cannot be granted on a product itself, but only on a process for making it. However, some of these countries (Greece is one) will grant a patent on a process for preparing the mixture by mere mechanical mixing. Such a process would not be patentable in the United States and in most countries where patents on processes involving only mechanical mixing are barred.

In the United States, if an invention relates to a new use of an old product it may be patentable as a process of treatment. For example, patents are granted on processes for treating plants to eliminate pests, for treating animals to cure disease, and even for treating human beings. Some countries allow, others bar patents of this type. And most also bar processes for treating human beings or animals. But there are exceptions. There is a recent Australian decision on a process for injecting a chemical into animals, prior to slaughtering, to tenderize the meat. This process was held patentable, because it had "economic value." Patents on this type of process are barred in Canada and the United Kingdom and in other countries linked to Great Britain, on the grounds that there is no "industrial" result. One patentee got around this bar in India on a process which actually involved injecting cows with a disinfectant to cure mastitis. He described the invention as a process for obtaining milk free from mastitis organisms.

Alternative Protection

A selection may have to be made between alternative forms of protection. For example, the Dutch Patent Office Appeal Department decided in 1962 that in the case of the antidiabetic compound tolbutamide, a patent could be obtained either for the preparation by known methods of the active pharmaceutical compound for this definite medicinal purpose or for the preparation of pharmaceutical mixtures including the active compound with carriers. The first type, the process patent, would not be infringed by somebody coming along later with a new process for making the compound. The second type, the mixture patent, would protect against somebody making the mixture, even if the active substance was made by a new process, but it might not cover the manufacture of the active substance in bulk. In such instances, the problem is to pick the alternative that will give the best commercial protection.

This same situation occurred in Canada on a slightly different basis. The lower court overruled the Patent Office and held that one patent could be obtained on the process for producing the new antidiabetic substance (Canada restricts protection to a chemically produced medicine or food to the process) and another patent on a mixture containing it. The Supreme Court overruled this decision, so that a new chemically produced substance intended for food or medicine can be protected only in terms of a process of making it.

Mixtures

One of the most common ways of getting around the limitations on claiming a new use of an old substance is to claim a mixture of the substance with a carrier. The mixture is thought of, for patentability purposes, as a pharmaceutical composition which never existed before and as possessing the unexpected properties of the active substance for the particular new use. There is wide variation from country to country as to how broadly mixtures like this can be patented. In Canada, patents are granted on this type of mixture, even if there is nothing special about the carrier. The Canadian courts have not yet ruled directly on the validity of such a claim, and many such patents are producing royalties.

Thus, broader effective coverage can be obtained than is possible in the United States, where there must be some unusual relationship between the carrier and the active substance or where the composition must have a special form—for example, a suppository.

Inventions in Special Fields Unpatentable

Different countries have different bars to patents on particular types of inventions. In Austria, inventions relating to edible salt, explosives, tobacco, and other items of government monopoly are not patentable. In Belgium and Great Britain, contraceptives are held unpatentable on moral grounds. But call them something else, and a patent may be available.

The main targets for these bars are foods and medicines. Most countries have some limitation to patents in these fields, tied up with the fear that products vital to life may become scarce or overpriced. This feeling, largely emotional, is good political ammunition and unfortunately seems to be spreading with the creep of socialism and nationalism. In Canada, for example, the report of the Restrictive Trade Practices Commission recently advocated the abolition of patents on drugs for reasons more emotional than real and without concern about the economic consequences of taking away the patent incentive. In other countries, there are reports of attempts to weaken the patent law on medicinals. Italy, however, after several years of refusing patents in the medicinal field, has a new patent law in draft form which would make it possible to obtain patents in this field. The drug industry there has become parasitical, and those doing research would like to restore a healthier climate. Some companies are now filing applications there in the hope that this law will become effective, validating their rights.

Patents are granted in most countries on process for making medicinal products. In some, mixtures intended for medicinal use are patentable and in others not. Intermediates leading to medicinals may be patented as products in Canada (whose law bars product patents on foods and medicines). In most countries chemical intermediates leading to medicinals can be covered only by process patents. In either case, a patent directed to a key intermediate may be useful in protecting an otherwise unpatentable medicinal end product. The only course, in this field, is to check with a specialist in each country to find out what the current situation is.

It is not always easy to find out how a particular country describes a "medicine." The Senate of the Patent Court in Germany, in 1962, held that even an agent that does not produce any immediate physiological effect on the human organism is a medicine. In that case, the agent was a laxative. The Germans also consider that oral agents for removing body odor are medicines. On the other hand, they do not class as medicines, agents for reducing the breeding activity of hens, radio-opaque material, adhesive plasters, and surgical dressings. A diagnosing means is not a medicine, even though the body functions cooperate in bringing about the diagnostic effect. Canada calls a general anesthetic a medicine, a local anesthetic not.

Proof of Properties Required

In the United States, particularly in medicines, the Patent Office requires extensive proof of utility and safety before it will grant a patent. Few foreign countries require these proofs. In some, an unsupported statement that a product has such-and-such properties is enough. For example, in Canada, new mixtures of two or more active substances alleged to have a synergistic relationship that cannot be predicted are regularly patented. No showing is usually called for other than an allegation in the patent application stating that the technical results claimed are real. There are no judgments of the courts to support or condemn such patents, and many are bringing royalties.

The main types of requirements and restrictions on patenting abroad are as follows:

If the invention is a new chemical substance, a few countries will grant a patent on the product itself. If it is a medicinal, the number of these countries is sharply reduced. Most will grant a patent on the process of making the substance. But, if the invention can be defined not as a chemical substance but something else, say a shaped article, it may qualify for a product patent. If the invention is a medicinal or other class that is barred, perhaps it can be described in terms of a use that is not barred or in terms of intermediates or processes for making it and then patented to give useful protection.

A "new use" may qualify for a patent on a process of treatment in which a chemical substance is applied to a substrate. If not, perhaps a mixture of the substance with a carrier may qualify as a patentable composition. If not, there are still a few countries in which a process of mixing can be covered.

Finally, the possibility of covering the invention in any jurisdiction should not be abandoned until it has been checked thoroughly with a patent attorney familiar with foreign practice. The development should be brainwashed in terms of all known ways of claiming the invention. Then, if the situation still looks hopeless, a specialist in the country where protection is desired should be consulted in case the invention is the "exception that tests the rule."

RECEIVED October 10, 1963.

INDEX

A

Abandonment 20, 59
Adrenalin case 102, 103
Adversary process 72
Aerosol Research Co. v. Scovill Mfg. Co. 10
Affidavits 54
Agawam Co. v. Jordan 18
Ajax Metal Co. v. Brady Brass Co... 85
Alloys 64, 109
Alternative protection 110
Altoona Publixs Theaters Inc. v. Tri-Ergon 22
Aluminum Co. of America v. Sperry Products, Inc. 7
American Cyanamid Co. case 104
Ames v. Lindstrom 50
Analogs, patentability of 73
Anticipation 65
Appeals 57, 69
Archer v. Papa 45
Art 59
Article patents 109
Artificial grouping 71
Aspirin cases 84, 104
Assignment 35, 55
Atlantic & Pacific Tea Co. case 66
  See also Great Atlantic & Pacific Tea Co.

B

Bac v. Loomis 42
Bancroft v. Watson 93
Becton-Dickinson & Co. v. R. T. Scherer Corp. 17
Benecke, Ex parte 49
Bergel, In re 12
Berkman and Berkman case 103
Best thought 22
Birmingham v. Randall 9
Blue, John Co. v. Dempster Mill Mfg. Co. 24
Boats, self-propelled 2

Bowers v. Woodman 32
Bradley, Judge 6
Brady, Ex parte 75
Brainstorming 17
Bremmer, In re 8
Bros Inc. v. Browning Mfg. Co..... 9
Brown, Ex parte 55
Brown v. Edeler 21
Burggraf-Bockelmann, John, and Strandskov, F. B., patent 95

C

Carrier plus chemical 87
Categories of patents 82
Central Farmers Fertilizer Co. 39
Chamberlain, G. D., Ex parte 88
Change of purpose 85
Chemical compounds 101
Chemical utility 23
Chemicals, foreign patent coverage.. 107
Claims 62
Clark Thread Co. v. Willimantic Linen Co. 33
Ciba case 108
Coca-Cola 28
Cochrane v. Deener 88
Coes, Loring, Jr., In re 11
"Composed of" 64
Compositions of matter 64, 82, 101
"Comprising" 64
Conception record 54
"Conception" 63
Conner v. Joris 47
"Consisting of" 64
Consolazio, W. V., patent 94
Constitutional grant 2
Continuation-in-part 67
Contributory infringement 90
Coolidge patent 64
Correction of errors 71
Corroboration 23, 42
Crystalline form patents 83

D

Davidson, In re 55
Davis v. Carrier 17
DDT 108
Dennis v. Pitner 103
Didusch, W. P., patent 92
Disclaimer 70
Disclosure 63
Diligence 42
Dinwiddie v. St. Louis & O'Fallon
 Coal Co. 35
Documents, formal 49
Dodson, In re 9
Dosselman and Neymann 60
Dr. Salisbury Laboratories v. I. D.
 Russell Co. 87
Druey and Schmidt, In re 10
Drugs 61, 111
du Pont de Nemours & Co. v. United
 States 28
Duddy v. Solomon 45
Dunmore 30
Dwight & Lloyds Sintering Co. v.
 Greennalt 20

E

Electric welding 66
Ellis, Carlton 60
Ellis, In re 60
"Embodiment" 63
"Essential" 63
Ether 2
Exclusive right 1
Extended technical arms 16
Extraction 109

F

Farbenfabriken of Elberfeld Co. v.
 Kuehmsted 84
Ferguson, A., Jr., patent 87
Field 54
Filing fee 49
Finish-remover 60
Fink patent 16
Finley, In re 11
Fisher, In re 84
Fitch, John 2
Fluorescents 109
Foods 111
Foreign patent coverage 107

Fridolph v. Bechik 45
Fromberg, Inc. v. Thornhill 87

G

Gaiser v. Linder 45
General Chemical Co. v. Standard
 Wholesale Phosphate & Acid
 Works 18
General Electric 101
Generic claim 62
Genus 62
Gill v. U.S. 34
Goodyear Co. v. Ray-O-Vac Co. .. 9
Graver Tank Mfg. Co. v. Linde Air
 Products 67
Gray patent 64
Gray, In re 51
Great Atlantic and Pacific Tea Co.
 v. Supermarket Equipment Corp. 7
 See also Atlantic & Pacific Tea Co.
Greif, Martin, patent 94
Grimme, In re 60
Gruskin, Benjamin, patent 87
Gustavson, Ex parte 55
Guth v. Minnesota Mining & Manu-
 facturing Co. 37
Gysin, Hans, and Knusli, Enrico,
 patent 91

H

Harding v. Steingiser & Salyer 44
Harrison and Packman, Ex parte.... 9
Hartop and Brandes, In re 9, 45
Hass, In re 11
Heinze, Ex parte 50
Henkel, Ex parte 11
Henze, In re 11
Herr, In re 11
Hessel, Ex parte 88
Heywood-Wakefield Co. v. Small.. 31
Hoffman, Felix, patent 84
Hofmann, Albert, and Troxler,
 Franz, patent 92
Homologs 10, 73, 77
Hopkins, Samuel 2
Hoskins v. General Electric 86
Hotchkiss v. Greenwood 9
Houghton v. United States 34
Huyl & Patterson v. McDowell Co.,
 Inc. 7

INDEX

I

"Illustrative"	63
Improvement	59
Incandescent Lamp case	60
Inducement to infringe	90
Infringement	60, 66
Interference	41, 57
chronological procedure	43
Interpretation	57, 65
Invalidity	68
See also Validity	
Invention	61
elements	42
in special fields	111
standard of	13
Inventor entity	24
Inventor, wrong	19
Inventorship	15
Intervening right	71
Isenstead v. Watson	9, 62
Isomers	73, 77

J

Jaffee and Ogden v. Kassley	9
Jepson, Ex parte	93
Joint invention	20
Joint inventors, order of names of	24
Jones, Kennedy, and Rotermund patent	66
Jungersen v. Ostby & Barton Co.	13

K

Kalo Inoculant Co. v. Funk Bros. Seed Co.	19
Kendall Co. v. Tetley Tea Co.	23
King case	102
Kleinman v. Betty Dain Creations	54
Koch v. Lieber	45
Krementz v. S. Cottle Co.	9
Krimmel, In re	9, 62
Kuehmsted v. Farbenfabriken of Elberfeld Co.	84, 101, 102
Kuhne Identifying Systems Inc. v. United States	19
Kyrides v. Bruson	45

L

Laboratory testing	44
Lambooy, In re	12
Lane & Bodley Co. v. Locke	30
Larsen, In re	12
Larsen Products Corp. v. Perfect Paint Products, Inc.	24
Legal requirements	1, 4
Levine v. United States	54
Levy	9
Linde Air Products v. Graver Tank Mfg. Co.	67
Livingston, Robert	2
Location of use	89
Lohr and Spurlin, In re	12
Lowell	30
Lungberg, In re	11
Lyon v. General Motors Corp.	89

M

Mahn v. Harwood	7
Manufacturers	94
Markush claims	70
Marsh Nichrome patent	65
Maytag Co. v. Murray Corp. of America	12
McClurg v. Kingsland	30
Meaning of chemical patents	57
Medicines	61, 111
foreign patent coverage	107
Mergenthaler v. Scudder	16
Metallurgy	64
Method	59
Microbiological products	99, 104
Miller, Ex parte	43, 50
Mills, In re Victor	12, 96
Mineral Separation v. Hyde	18
Misjoinder, hazards of	24
Mixed Crystals	109
Mixtures	109, 110
Moler & Adams v. Purdy	43
Monsanto Chemical Co. v. Miller	38
Morehouse, N. F., & Mayfield, O. J., patent	87
Morse case	100
Morton, Dr.	2
Morway, Beerbower & Zimmer v. Bondi	42
"Multiplicity"	63

N

Natural products	99
Nelson and Shabika case	8, 61

New matter 67
New use 11
Nichols v. Atkinson 9
Nielsen v. Cahill 21
Nitrogen-fixing bugs 19
Nonjoinder, hazards of 24
Nonpatentable discovery 2
Novak and Hogue, In re 9
Novelty 4, 78

Process claims 88
patents 59, 108
Proof of utility 44, 112
Prusak 9
Purity alone 85
Pyrene Mfg. Co. v. Boyce 33
Pyrophoric alloy 62

R

Radio Position Finding Corp. v. Bendix Corp. 81
Randolph, Edmund 57
Records 53
Reduction to practice 42, 44
Rehearing 70
Reifsnyder, Ex parte 55
Reiner v. The I. Leon Co., Inc. .. 13
Reissue 70
Restrictive Trade Practices Commission 111
Rhodes, Ex parte 49
Riden and Flavin, In re 11, 95
Rodin v. Spalding 44
Rothermel and Waddell, Jr., In re.. 10
Rule 71 61
Rule 75 ....62, 68
Rules 131 and 132 54
Rules of general application 96
Rules of practice 63
Rystan Co. v. Warren-Teed Products Co., Inc. 87

O

Oath 49, 50
Objects of the invention 66
Obviousness 74
  See also Unobviousness
Official rules 58
O'Reilly v. Morse 18, 100
Ownership 27, 32

P

Palmquist and Erwin, In re 9
Papers, formal 49
Papesch, In re Viktor 8, 75, 82
Parke-Davis v. Mulford 102
Patent Acts 2
Patent Code of 1952 ....70, 106
Patent Law, First U.S. 29
Patentability 7
Patenting abroad 112
Patents, first American 2, 57
Pearl ash 2, 58
Petering and Fall, In re 12
Petition 49, 52
Pfeiffer, F. L., patent 83
Pharmaceutical applications 62
Phillips Petroleum Co. v. Ladd ... 94
Pieroh and Weeres, In re 11, 88
Pigments 109
Plant isolates 99, 103
"Plurality" 63
Pointer v. Six Wheel Corp. 19
Potash and pearl ash 2
Power of attorney 51
"Preferably" 60, 63
Preliminary statement 53
  amendment 43
Principle of nature 99
Prior art references 12
Priority of invention 42
Products of nature 99

S

Salathiel, Ex parte 93
Salt 2
Santmyer, Ex parte 88
Schaefer, Inc., v. Mohawk Cabinet Co., Inc. 13
Scherer, Ex parte 90
Schluchter, Ex parte 55
Secrecy provision 38
Section 33 68
Section 101 .... 8, 61
Section 100(b) 65
Section 103 .... 9, 59
Section 112 61
Semiconductors 109
Shell Development v. Watson 86
Shop right 29
Short form 53

INDEX

Silicates	68
Six-wheel truck	19
Smith Brothers cough drops	28
Snell, Ex parte	103
Solomons v. United States	30
Solutions	109
Special purity	84
Species	62
"Specific embodiment"	59
Standard Parts Co. v. Peck	33
Statutory changes	81
Steelmand, Ex parte, S. L., and Kell, T. L.	85
Steinberg, M. D., patent	92
Steroid	61
Structure in compositions	94
"Subject matter as a whole"	74
"Subject matter sought to be patented"	74
"Substantially"	63
Symington Co. v. National Castings Co.	33
Szwarc, In re	8

T

Tanczyn, In re	85
Tansel, In re	17
Tea bag container	23
Teppema, Ex parte	55
Tetracycline case	104
Thropp & Son v. DeLaski and Thropp Circular Woven Tire Co.	21, 24
Thuau, In re	96, 104
Tracerlab Inc. v. Industrial Nucleonics Corp.	16
Trade secrets	28, 36, 38
Treatment of animals	91
Treatment of human body	91
Tolbutamide	110
Tolkmith	61
Twentier's Research Inc. v. Hollister Inc.	7
Tungsten, metallurgy of	64
Tungsten case	101

U

Udy patent	16
Union Carbide case	66

United Chromium v. General Motors Corp.	16
United Shirt & Collar Co. v. Beattie	22
United States v. Dubilier Condenser Corp.	29
United States v. Solomons	34
Unobviousness	5, 9, 59, 106
See also Obviousness	
Use to be claimed as process	88
Utility	5, 23, 61, 105
contemplated	44
proof of	44, 112
requirements of	8
ultimate use doctrine	45

V

Validity	7, 66
See also Invalidity	
Van Otteren v. Hafner	24
Visible fingerprints	19
Von Bramer & Ruggles, In re	95
Vrooman v. Penhallow	22

W

Warp v. Warp	60
Washington, George	57
Welsbach patent	62
Westinghouse Electric Co. v. Montgomery	68
Wever v. Good & Putzrath	42
Wickwire Spencer Steel Co.	32
Widmann, R. R., patent	87
Wilke and Pfohl, In re	8
Willard, J. R., and Maitlen, E. G.	92
Winslow, Samuel	2
Wiswall, R. H., Jr., patent	83
Writ of *certiorari*	69
Wrong inventor named	19

Y

Yale and Bernstein, Ex parte	84

Z

Zbornik and Peterson, Ex parte	91